T0135706

Analysis of the mechanical performance of pin-reinforced sandwich structures

Vom Fachbereich Produktionstechnik
der
UNIVERSITÄT Bremen

Zur Erlangung des Grades
Doktor-Ingenieur
genehmigte

Dissertation

von
M.Sc. Mohamed Adli Dimassi

Gutachter: Univ.-Prof. Dr.-Ing. Axel S. Herrmann, Universität Bremen
Univ.-Prof. Dr.-Ing. habil. Lothar Kroll, Technische Universität Chemnitz

Tag der mündlichen Prüfung: 12. August 2019

Science-Report aus dem Faserinstitut Bremen

Hrsg.: Prof. Dr.-Ing. Axel S. Herrmann

ISSN 1611-3861

Bibliographic information published by Die Deutsche Bibliothek

Die Deutsche Bibliothek lists this publication in the Deutsche Nationalbibliografie; detailed bibliographic data is available in the Internet at http://dnb.ddb.de.

ISBN 978-3-8325-5010-3

Logos Verlag Berlin GmbH
Comeniushof, Gubener Str. 47,
10243 Berlin
Tel.: +49 030 42 85 10 90
Fax: +49 030 42 85 10 92
INTERNET: http://www.logos-verlag.de

Acknowledgements

This work was performed during my employment as scientific assistant at the Faserinstitut Bremen e.V. (FIBRE). I would like to thank my colleagues for their numerous discussions and help, and specially Dr.-Ing. Christian Brauner for his support and valuable advices. This work was carried within the research project iSand (grant ID 20W1306D), which is funded by the German Federal Ministry for Economic Affairs and Energy through the federal Aeronautical Research Program V-1 (LuFo V-1). At this point I want to gratefully acknowledge this support and to thank all project partners for the good and successful collaboration. Furthermore, I would like to thank the European Regional Development Fund (2007-2013) and the Wirtschaftsförderung Bremen (WFB-QS 1005) for enabling access to the micro-CT equipment at the University of Bremen.

I would like to express my gratitude to my first supervisor Prof. Dr.-Ing. Axel S. Herrmann for giving me the opportunity to develop this PhD thesis and for providing all the necessary support and infrastructure as well as his constant encouragement during my research activities at FIBRE. In addition, I would like to thank Univ.-Prof. Dr.-Ing. habil. Lothar Kroll for accepting the position of second supervisor of this thesis. I would like also to thank all members of the doctoral committee for their presence and engagement in the PhD colloquium.

Finally, I would like to thank my parents Moncef and Samia for their sacrifices, support and for giving me the opportunity to study in Germany through their financial support. Moreover, I would like to thank my wife Yesmine for always encouraging me and supporting me especially during difficult moments. Special thanks go to my uncle Zouhir, my parents-in-law Youssef and Rafika as well as my grandparents for their continuous support and encouragement.

Abstract

In the last decade the strong rise in oil prices has led to the increase of the needs to lighter aircrafts with efficient engines to reduce the kerosene consumption and the operating costs. Composite materials with their high specific strength and stiffness are widely used in the latest aircraft designs, but their intensive use is restricted to long-range aircrafts due to high manufacturing costs and low production rate. Integral designs and using dry fibre preforms overcome this drawback. Foam core sandwich structures combine the advantages of high bending properties with low manufacturing costs when liquid composite processes are used. Moreover, the high buckling stability of sandwich structures leads to the discard of stiffeners and thus the reduction of manufacturing steps. However, the use of foam core sandwich structures is not widespread in aircraft applications due to the better weight-specific performance of honeycomb cores and the susceptibility to impact loading. In this context, pin reinforcements are added to the foam core to improve its mechanical properties and its damage tolerance.

In order to enable the use of pin reinforced foam core in a primary aircraft structure, it is important to know its behaviour at different loading conditions and to predict its impact behaviour. This thesis contributes to the understanding of the damage mechanisms of pin reinforced foam core sandwich structures under different loading conditions. Quasi-static tests were performed and the micro-computed tomography was used to understand the damage occurrence under compressive, shear and indentation loads. The pins led generally to the improvement of the mechanical performance under static loading conditions. It was found out that a minimum pin volume fraction in the foam core is necessary to reach a remarkable improvement of the mechanical properties.

Since loading-carrying aircraft structures operate generally in extreme frost conditions, impact tests on pin reinforced foam core sandwich structures were performed at -55 °C and the damage behaviour was compared to the damage in panels impacted at room temperature. It was concluded that the residual stresses created in the vicinity of the carbon fibre pins during the manufacturing process and due to the thermal loading in operating condition lead to the degradation of the impact performance and create thermal cracks in the foam material. These cracks can dispread under further loading and endanger the integrity of the structure. Reducing the thermal stresses in the foam core by using fibre with high coefficient of thermal expansion like glass fibre or by modification of manufacturing parameters led to the delay of the thermal stresses to higher impact energies.

An FE-model simulating the impact behaviour of pin reinforced foam core sandwich panels was developed by using the simulation program Abaqus and following the building block approach. Two approaches to model the pin reinforced foam core were investigated and assessed. The homogenisation approach delivered more reliable results than the approach with discrete pin modelling. The simulation results were validated by using impact test results and an overall good agreement was achieved. The proposed model allows predicting the impact behaviour of sandwich structures and is able to simulate impact on structures with complex geometrical features so that it can be used for design studies and thus reduces the testing effort. A further improvement of the simulation results can be achieved by considering the residual stresses due to manufacturing and thermal loading in the foam, and by considering the damage of the pins if the pins are modelled.

Keywords: composite, impact, pin reinforced foam core, Tied-Foam-Core, sandwich, explicit FEM, micro-computed tomography.

Kurzfassung

In den letzten zehn Jahren hat der starke Anstieg der Ölpreise zur Verstärkung des Bedarfs an leichteren Flugzeugen mit effizienten Triebwerken geführt, um den Kerosinverbrauch und die Betriebskosten zu senken. Verbundwerkstoffe mit ihrer hohen spezifischen Festigkeit und Steifigkeit sind in den neuesten Flugzeugmodellen weit verbreitet, ihr intensiver Einsatz beschränkt sich jedoch aufgrund hoher Herstellungskosten und niedriger Produktionsraten auf Langstreckenflugzeuge. Integrale Designs und die Verwendung von Preforms aus Trockenfasern überwinden diesen Nachteil. Schaumkern-Sandwichstrukturen kombinieren die Vorteile hoher Biegeeigenschaften mit niedrigen Herstellungskosten beim Einsatz von Liquid-Composite-Moulding-Fertigungsverfahren. Darüber hinaus führt die hohe Knickstabilität von Sandwichstrukturen zum Wegfall von Versteifungen der Composite Schale und damit zur Reduzierung von Fertigungsschritten. Der Einsatz von Schaumkern-Sandwichstrukturen ist jedoch in Flugzeuganwendungen aufgrund der besseren gewichtsspezifischen Leistung der Wabenkerne und der Anfälligkeit für Stoßbelastungen nicht weit verbreitet. In diesem Zusammenhang werden dem Schaumkern Pin-Verstärkungen hinzugefügt, um seine mechanischen Eigenschaften und seine Schadenstoleranz zu verbessern.

Um den Einsatz von Pin-verstärktem Schaumkern in einer primären Flugzeugstruktur zu ermöglichen, ist es wichtig, sein Verhalten bei unterschiedlichen Lastbedingungen zu kennen und sein Aufprallverhalten vorherzusagen. Diese Arbeit trägt zum Verständnis der Schadensmechanismen von Pin-verstärkten Schaumkern-Sandwichstrukturen unter verschiedenen Lastbedingungen bei. Quasi-statische Tests wurden durchgeführt und die Mikro-Computertomographie wurde eingesetzt, um das Schadensbild unter Druck-, Scher- und Eindruckbelastung zu verstehen. Die Pins führten im Allgemeinen zur Verbesserung der mechanischen Leistung unter statischen Lastbedingungen. Es wurde festgestellt, dass ein minimaler Pinvolumenanteil im Schaumkern notwendig ist, um eine deutliche Verbesserung der mechanischen Eigenschaften zu erreichen.

Da tragende Flugzeugstrukturen im Allgemeinen unter extremen Gefrierbedingungen arbeiten, wurden bei -55 °C Schlagversuche an Pin-verstärkten Schaumkern-Sandwichstrukturen durchgeführt und das Schadensverhalten mit dem Schaden von bei Raumtemperatur beaufschlagten Platten verglichen. Es wurde festgestellt, dass die während des Herstellungsprozesses in der Nähe der Kohlefaserpins erzeugten Eigenspannungen und aufgrund der thermischen Belastung im Betriebszustand zu einer Verschlechterung der Impactleistung führen und thermische Risse im Schaumstoff erzeugen. Diese Risse können sich unter weiterer Belastung ausbreiten und gefährden die Integrität der Struktur. Die Reduzierung der thermischen Spannungen im Schaumkern durch die Verwendung von Fasern mit hohen Wärmeausdehnungskoeffizienten wie Glasfasern oder durch Änderung der Herstellungsprozessparameter führte zur Verzögerung der thermisch induzierten Risse auf höhere Schlagenergien.

Ein FE-Modell zur Simulation des Aufprallverhaltens von Sandwichpaneelen mit Pin-verstärktem Schaumkern wurde mit dem Simulationsprogramm Abaqus und nach dem Baukastenprinzip entwickelt. Zwei Ansätze zur Modellierung des Pin-verstärkten Schaumkerns wurden untersucht und bewertet. Der Homogenisierungsansatz lieferte zuverlässigere Ergebnisse als der Ansatz mit diskreter Pin-Modellierung. Die Simulationsergebnisse wurden durch die Verwendung von Impact-Testergebnissen validiert und eine insgesamt gute Übereinstimmung erzielt. Das vorgeschlagene Modell ermöglicht die Vorhersage des Impactverhaltens von Sandwichstrukturen und kann die Strukturantwort von Strukturen mit komplexer Geometrie simulieren, so dass es für Designstudien verwendet werden kann und somit den Testaufwand reduziert. Eine weitere Verbesserung der Simulationsergebnisse kann durch die Berücksichtigung der Eigenspannungen im Schaum, die durch Fertigung und thermische Belastung erzeugt wurden, sowie durch die Berücksichtigung der Beschädigung der Pins bei dessen Modellierung erreicht werden.

Schlüsselwörter: Faserverbundstrukturen, Pin-Verstärkung, Sandwichstruktur, Impact, explizite FEM, Mikro-Computertomographie, Tied-Foam-Core.

Publications

Parts of this PhD project have been published in the following publications and conference presentations:

Dimassi MA, Brauner C and Herrmann AS. Experimental study of the mechanical behaviour of pin reinforced foam core sandwich materials under shear load, In: Proceedings of the 18th Chemnitz seminar on materials engineering, Chemnitz, Germany, 2016; 59, pp. 268-275.

Dimassi MA, Brauner C, Herrmann AS, et al. Experimental study of the indentation behaviour of tied foam core sandwich structures, In: Proceedings of the 17th European Conference on Composite Materials, Munich, Germany, 2016.

Dimassi MA and Herrmann AS. Numerical Simulation of low velocity impact on pin-reinforced foam core sandwich panel. 21. Symposium Verbundwerkstoffe und Werkstoffverbunde, Bremen, Germany, July 2017.

Dimassi MA, Brauner C, Focke O et al. Experimental investigation of Tied foam Core sandwich compression performance. J. Sandwich Struct. Mater. 2017; DOI: 10.1177/1099636217748577

Dimassi MA, John M and Herrmann AS. Investigation of the temperature dependent impact behaviour of pin reinforced foam core sandwich structures. Compos. Struct. 2018; 202, pp. 774-782.

Table of contents

X

Catalogue of symbols used

Abbreviations, Definitions

3PB	Three Points Bending
Al	Aluminum
BBA	Building Block Approach
CAD	Computer-aided Design
CAI	Compression After Impact
CFL	Courant-Friedrichs-Lewy condition
CFRP	Carbon Fibre Reinforced Polymer
CT	Computed Tomography
CTE	Coefficient of Thermal Expansion
CZM	Cohesive Zone Model
DCB	Double Cantilever Beam
ENF	End Notched Flexure
EP	Epoxy Resin
FE	Finite Element
FEM	Finite Element Method
FRP	Fibre Reinforced Polymer
FST	Fire, Smoke and Toxicity
GFRP	Glass Fibre Reinforced Polymer
HTS	High Strength fibres
IMWS	Institute for Microstructure of Materials and Systems
LCM	Liquid Composite Moulding
LEFM	Linear Elastic Fracture Mechanic
MMB	Mixed Mode Bending
NCF	Non-Crimp Fabrics
NDE	Non-destructive Evaluation
NDT	Non-destructive Testing
PEEK	Polyether Ether Ketone
PES	Polyester
PH	Phenolic Resin
PMI	Polymethacrylimide
Prepreg	Preimpregnated Fibres
PPS	Polyphenylene Sulfide
PS	Polystyrene
PU, PUR	Polyurethane
PVC	Polyvinychloride
PVF	Pin Volume Fraction
RTM	Resin Transfer Moulding
SAN	Styrene Acrylonitrile
SEM	Scanning Electron Microscope
SHPB	Spit Hopkinson Pressure Bar
SiC	Silicon Carbide
TFC	Tied Foam Core
UD	Unidirectional
US	Ultrasonic

VARI	Vacuum Assited Resin Infusion
VARTM	Vacuum Assisted Resin Transfer Moulding
VCCT	Virtual Crack Closure Technique
VE	Vinil Ester Resin

Notations

A_C	Initial cross section surface of test specimen for compression test
A_L	Minimum sufficient area for load introduction
a	Radius of the crushing zone
$\bar{a}$	Dimensionless radius of the crushing zone
b	Width of a sandwich beam
C_d	Damaged elasticity matrix
c_d	Dilatational wave speed of the material
c^M	Correction to the solution of the Taylor series expansion at degree of freedom
D	Beam bending stiffness
D_0	Bending stiffness of the face sheets about the neutral axes
D_c	Bending stiffness of the core
D_f	Bending stiffness of the face sheet
d_f, d_m, d_s	Internal damage variables for fibre, matrix and shear damage
E	Beam elastic modulus
E_1, E_2	Young´s Modulus in the fibres direction and Young´s modulus perpendicular to the fibers in a composite UD-layer
E_{33}	Young´s Modulus in thickness direction of the laminate
E_c	Elastic modulus of the core
E_{cr}	Absorbed crushing energy of the core
E_{cz}	Compressive modulus of the foam core
E_{czm}	Cohesive zone stiffness
E_f	Elastic modulus of the face sheets in a sandwich construction with similar layup
E_{f1}, E_{f2}	Elastic moduli of the face sheets in a sandwich construction with different layups
E_{fc}	Compressive modulus of the foam core
E_p	Compressive modulus of the pin
$E_{p,x}, E_{p,y}$	E-moduli of a pin along and transverse to the pin axis
E_r	Average Young´s modulus of a composite skin
F	Contact load during indentation and impact
F_B	Bending load acting on the pin
F_{BS}	Load at shear damage during sandwich shear test
F_c	Compressive load acting on the pin
F_{cr}	Critical load for core crushing
F_{CS}	Critical load for core shear cracking
F_{d1}	Critical load for skin delamination
F_f^t, F_f^c	Damage onset factors of the fibre under tensile and compression loading
F_L	Punctual load
F_m^t, F_m^c	Damage onset factors of the matrix under tensile and compression loading
F_N	Normal contact load

F_p	Pin load
F_R	Friction load
F_r	Skin rupture load
f_{fc}	Volume fraction of foam material in a pin-reinforced foam core sandwich
f_p	Volume fraction of the pins in a pin-reinforced foam core sandwich
G_c	Shear modulus of the core
G_{ce}	Critical damage dissipation energy
G_x^T, G_x^C	Fracture energies for longitudinal tension and compression damage of a composite UD-layer
G_y^T, G_y^C	Fracture energies for transverse tension and compression damage of a composite UD-layer
$G, \vartheta_{12}, \vartheta_{21}$	Shear modulus and the Poisson's ratios of a composite UD-layer
G_{Ic}, G_{IIc}	Critical energy release rate for mode I and mode II
h	Height of the beam
h_c	Thickness of the core
$h_{c,min}$	Minimum core thickness to avoid core shear cracking
h_{cz}	Thickness of the sublaminate
h_d	Distance between the centroids of the face sheets
h_f	Thickness of the face sheet
I	Second moment of areal inertia of the beam
i	Iteration step
k	Compression yield stress ratio
k_t	Hydrostatic yield stress ratio
K_{cz}	Cohesive zone stiffness
K_n, K_s, K_t	Normal and shear cohesive zone stiffnesses
K^{NM}	Jacobean matrix
L_c	Characteristic length for damage evolution
L^e	Characteristic length of an element
l_{cz-I}, l_{cz-II}	Cohesive zone length for mode I and mode II
l_e	Length of the cohesive element
l_1	Length of the shear test specimen
M	Mass of the impactor
M_{cz}	Cohesive zone model variable
M_x	Bending moment about x-axis
M_y	Bending moment about y-axis
M^{NM}	Mass matrix of FE-model with subscript N as degree of freedom of the model
m_p	Mass of the impacted panel
N	Skin membrane load
N_e	Number of cohesive elements within a cohesive zone
N_x	In-plane load
P	Compressive load
P_{SC}	Critical shear crimping load
P^N, I^N	External and Internal force vectors created by stress in an element with subscript N as degree of freedom of the model
p, q, S	Pressure stress, Mises stress and deviatoric stress of a foam material
p_c	Yield stress in hydrostatic compression
p_{cr}	Crushing stress of the core

p_c^0	Hydrostatic compressive strength
p_t	Hydrostatic tensile strength
Q_{cz}^*	Effective stiffness of the core
Q_f^*	Effective stiffness of the face sheet
R_{imp}	Radius of the hemispherical impactor
S^a	Artificial shear strength of the cohesive zone
S^L, S^T	Longitudinal and transverse shears strength of a composite UD-layer
T_x	Equivalent transverse force
T_0, S_0	Tensile and shear strengths of the resin
T^a	Artificial normal strength of the cohesive zone
t	Impact duration
t_n^0	Nominal stress for purely normal deformation of a cohesive element
t_s^0, t_t^0	Nominal stress for purely shear deformation of a cohesive element in first or second shear direction
t_n, t_s, t_t	Stress components of the cohesive zone elements after damage initiation
$\bar{t}_n, \bar{t}_s, \bar{t}_t$	Predicted stress components of the cohesive zone elements using the linear elastic traction-separation low for the current strain without damage
$u^N, \dot{u}^N, \ddot{u}^M$	Displacement, velocity and acceleration vectors
V	Current volume of an FE-model
v	Impact velocity
X^T, X^C	Longitudinal tension and compression strengths of a composite UD-layer
Y^T, Y^C	Transverse tension and compression strengths of a composite UD-layer
ρ	Beam density
ε_c	Normal strain in the core
ε_{cz}	Compressive strain
ε_f	Normal strain in the skins
ε_{x0}	Normal strain at the neutral axis of the sandwich beam
ε^f	Fracture strain
ε_{pl}^{vol}	Volumetric plastic strain
ε_{1t}	Ultimate tensile failure strain of UD-layer in fibre direction
$\dot{\varepsilon}$	Strain rate
σ_c	Normal stresses in the core
$\hat{\sigma}_c$	Compressive strength of the core in loading direction
$\bar{\sigma}_{cz}$	Compressive strength of the foam core in z-direction
σ_{cz}	Compressive stress
σ_f	Normal stresses in the skins
σ_{fc}	Compressive strength of the foam material
σ_{f1}, σ_{f2}	Stresses in different face sheets due to in-plane load
$\hat{\sigma}_{f,FW}$	Critical wrinkling stress for sandwich beam
$\sigma_p(\phi)$	Compressive strength of a pin for an inclination angle ϕ
$\bar{\sigma}_{11}, \bar{\sigma}_{22}, \bar{\tau}_{12}$	Components of the effective stress tensor $\bar{\sigma}$ that considers the actual damage situation in a composite UD-layer
τ_c	Shear stress in the core
$\bar{\tau}_c$	Shear strength of the core
$\tau_{c,max}$	Maximum achievable shear stress in the core during impact
$\hat{\tau}_{c,x}, \hat{\tau}_{c,y}$	Allowable shear stresses of the core in x and y direction
$\bar{\tau}_{c,xz}, \bar{\tau}_{c,yz}$	Maximum core shear stresses in xz- and yz-plane

τ_f	Shear stress in the face sheets
$\tau_{p,xy}$	Shear stress in a pin
α	Shape factor of the yield surface of the volumetric hardening foam model
α_{cz}	Cohesive zone parameter
α_{HS}	Contribution factor of the shear stress to the fibre tensile failure onset
δ_m	Mixed-mode displacement
δ_m^0	Effective displacement at damage initiation a the cohesive element
δ_m^f	Effective displacement at complete failure of the cohesive element
δ_m^{max}	Maximum value of effective displacement achieved during load history of the cohesive element
$\delta_n, \delta_s, \delta_t$	Normal and shear deformations across the cohesive interface
$\vartheta_{p,xy}$, $G_{p,xy}$	Poisson´s ratio and shear modulus of a pin
ϑ_r	Average Poisson´s ratio of the skin
η_{BK}	Benzeggagh-Kenane parameter
β^N	Strain rate-displacement rate transformation
μ	Friction coefficient
λ, μ_l	Lame´s constant
θ	Angle between the membrane load direction and the horizontal plane
ϕ	Inclination angle of the pins
φ	Fibre volume content
Δt	Stable time increment for explicit simulation
Δh	Crosshead displacement during compression test

1 Introduction

1.1 Motivation

Due to their high specific strength and stiffness Fibre Reinforced Polymer (FRP) composites are widely used where structures with high mechanical performance combined with lightweight are needed, e.g. in the maritime, wind energy and aerospace industries. In the last decade the use of high performance carbon fibre reinforced composite materials has proven its effectiveness in the aerospace industry. Composite materials have even substituted most of the metallic structures in the latest civil aircrafts to reach more than 50% of the structural weight in the Boeing 787-Dreamliner and the Airbus 350 XWB. In addition to weight reduction, composite materials offer many advantages like environmental durability, superior fatigue properties, life cycle costs and service life extension. However, due to the complex mechanical behaviour and using conventional design approaches that are based on metallic design approaches, manufacturing costs are very high compared to metal manufacturing processes and the full potential of these materials has not been used yet. The high materials and manufacturing costs limit the use of composite materials to high-end products with strict lightweight requirements.

The combination of innovative composite design with efficient manufacturing techniques would lead to a better exploitation of the potential of composite materials and to the reduction of manufacturing costs. Integral manufacturing of composite parts improves the mechanical performance of the structure and reduces the material and manufacturing costs as the parts count and the assembly time are extremely reduced. Sandwich composite structures consisting of two strong and stiff composite face sheets bonded to a rigid core offer the possibility to manufacture integral composite parts while having good mechanical performance. Sandwich structures with a closed cell foam core offer many advantages compared to stiffened monolithic shells and to the in aircrafts widely used honeycomb sandwich. It combines the high specific bending stiffness and superior buckling stability [1] with the possibility to manufacture sandwich panels with complex geometries using cost effective resin infusion technologies [2]. Moreover, closed cell foam core overcomes the issue of moisture take up of honeycomb sandwich, has excellent acoustic damping and thermal insulation properties, which makes the foam core sandwich attractive for the integration in primary aircraft structure. In addition, foam core material has good energy-absorbing properties with nearly a constant crushing load level [3], which makes it interesting for crash absorbing applications in automotive and aerospace industry [4-6]. However, sandwich structures are susceptible to external localised loads normal to the face sheets, like low-velocity impact and inappropriate load introduction. Such kind of loading could lead to face sheet rupture, local core crushing, interface debonding and microcracks, as well as in worst case shear cracks [7]. These failures could degrade the load bearing capacity of the panel, propagate under further loading and threaten the integrity of the structure.

In order to enhance the compressive properties of foam core sandwich and its damage tolerance, many innovative through the thickness reinforcement solutions like pin reinforcement [8, 9], double-T reinforcement [10], hierarchical and foam filled corrugation structure [11] and stitch bonded sandwich structure [12] have been proposed. In order to enable the use of this kind of hybrid sandwich composites in primary aircraft structure, it is essential to understand and assess the damage behaviour at different loading conditions

1

and to provide reliable simulation methods to design the structure. On the one hand, the experimental investigation provides an understanding of the mechanical behaviour and the damage modes, which enables to know the limits of the used materials. On the other hand, simulation methods enable a faster structure development with reduced scrap rates and engineering effort.

1.2   Objective

The first main objective of this PhD thesis is to investigate the influence of pins on the mechanical performance and the damage behaviour of pin-reinforced foam core sandwich structures. The pins are embedded in the foam core under a specific inclination angle and predefined pattern using the Tied Foam Core (TFC) pinning technology. Different quasi-static tests were performed, namely flatwise compression, shear and indentation tests. X-ray computed tomography was used as well to identify the damage modes under static loading. The main effects and influencing parameters should be detected and the gathered findings should be compared to the published results of similar pin-reinforced foam core sandwich structures; differences and similarities should be identified.

The second objective is to investigate the temperature dependent impact behaviour of the pin-reinforced foam core sandwich structure. Low-velocity impact tests were performed at room temperature and very low temperature (-55 °C). The damage behaviour was investigated by using ultrasonic C-scan and X-ray computed tomography. In this work the influence of the very severe frost conditions on the impact damage occurrence should be thoroughly investigated and analysed.

The last main objective of this work is to develop a simulation model to describe the damage behaviour of the studied structure under low-velocity impact. The model should be capable of predicting the different damage modes and the load-displacement response with good accuracy, so that it can be applied to large sandwich structures with real boundary conditions.

Finally, it is expected that this PhD thesis will contribute to a significant understanding of pin-reinforced foam core sandwich structures, which would widespread the use of these materials in aerospace applications.

1.3   Structure of the work

This PhD Thesis consists of six chapters. In this chapter an introduction into the work including its motivation, objective and structure is provided.

The second chapter starts with a general overview about the state of the art of sandwich structures, its applications and manufacturing technology as well as the fundamentals of the sandwich theory and standard damage modes. Then the failure modes under low-velocity impact are explained and an analytical model to predict the impact behaviour of foam core sandwich structures is presented. The chapter is concluded with a literature review about through-the-thickness reinforcement of composite sandwich structures and a detailed presentation of the used pinning technology (TFC-technology).

The third chapter focuses on the characterisation of the used pin-reinforced foam core. Different quasi-static tests were performed in which four pin configurations were investigated. Concerning the compressive properties, an analytical model to predict the compressive strength and stiffness was assessed and applied to the tested materials. In addition to visual inspections X-ray computed tomography was used to determine the damage modes at every loading condition investigated in this work.

The following chapter four covers the impact response of TFC-sandwich at low temperature and discuss the influence of very low-temperature on the damage behaviour. Test procedure and results for impact tests at room temperature and -55 °C are presented. The damage modes were investigated using ultrasonic C-scan and X-ray computed tomography. An investigation of the influence of the manufacturing process and the pin material on the damage mode at -55 °C is provided and the results are analysed.

In chapter five, the development of a numerical model to simulate the impact behaviour of pin-reinforced foam core sandwich structure is presented. In a first part the fundamentals of explicit simulation are given and the used assumptions, material models and modelling approach are explained. In the second part, a numerical model for impact on foam core composite sandwich structure, which is the basic of the model of the pin-reinforced foam core sandwich, is introduced. The last part of the chapter deals with the simulation of the low-velocity impact on pin-reinforced foam core sandwich structure. All simulation results were compared to the experimental results for validation.

In the last chapter, the results are summarised, limitations and recommendations for an efficient implementation of the results are given and prospects to improve the presented findings of this work are proposed.

2 State of the art

Chapter 2 is giving an overview about the state of the art of composite sandwich structures. In the first part an insight into the fundamentals of composite sandwich structures is presented with focus on applications, material combination and manufacturing. In the second part, the low-velocity impact behaviour of foam core sandwich structures is considered, where the impact failure modes and the analytical method to predict the impact response of foam core sandwich structures are discussed. The third section provides a critical review about available through-the-thickness reinforcement methods of foam core sandwich structures. The motivation behind using the through-the-thickness reinforcement, the properties and the manufacturing processes are highlighted. The impact behaviour of foam core sandwich structures and the effects of through-the-thickness reinforcement on the mechanical properties of sandwich composites provide the basis of the performed research work. Lastly, the tied foam core (TFC) pinning technology investigated in this work is thoroughly described and its advantages compared to other technologies are outlined.

2.1 Fundamentals of composite sandwich structures

2.1.1 Definition

There are different tales about the origin of the word sandwich, which describes a tasty snack consisting of two slices of bread with a filling of e.g. meat and cheese. One of these tales recounts that John Montagu (1718-1792) the 4th earl of Sandwich and British first lord of the Admiralty during the American Revolution gave his name to this dish. He was a devoted Cribbage-player and spent hours playing at his table without any food except his sandwiches. Sandwich structures are built in similar manner to the sandwich dish, the two bread slices are replaced by two skins of high performance materials like metal sheets or composites and the filling is replaced by a hard lightweight core [13].

Industrially used sandwich structures have three main components (Figure 1): two thin, strong and stiff skins separated by a thick, hard and lightweight core with negligible strength compared to the face sheet material. Both components are bonded to each other by adhesive medium to ensure structural integrity and load transfer. The sandwich functional mode is similar to that of an I-beam with more material efficiency, as different materials are used and the mechanical potential of the face sheets is better exploited. The rigid core maintains the distance between the skins constant, resists shear and out-of-plane compression and provides support to face sheets, which increases the resistance against bending and buckling. The face sheets work together and resist in-plane loads, namely tensile and compression stresses resulted from bending loads. The adhesive layers bond the face sheets and the core together and ensure the load transfer between them. As the sandwich concept works only when all elements are bonded together it is important that the adhesive joint has enough strength to withstand the shear and tensile stresses set up between the skins and the core. A huge variety of materials exists for the sandwich elements, which makes the design of sandwich structures very time consuming, but it offers the possibility to tailor the mechanical properties of the sandwich to the specific application. Moreover, every material combination should separately be investigated as the damage behaviour and the mechanical properties depend strongly on the used sandwich constituents and the manufacturing process.

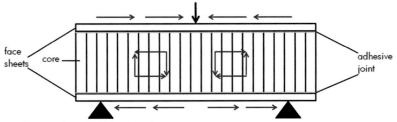

Figure 1. Main elements of a sandwich construction

2.1.2 Application of composite sandwich structures

The first discussion about the advantages of sandwich constructions is associated to Duleau who, in 1820, tested the bending performance of two beams assembled with a constant distance to each other. He concluded that the bending stiffness increases by the third power of the distance between the beams [14]. The sandwich concept was firstly adopted by Fairbairn, in 1849, who built bridges with sandwich elements made of steel sheets and corrugated steel core [15]. Due to the lack of high performance materials and suited structural adhesive, it took 110 years to use the concept of sandwich structure commercially [13]. The earliest serial application of composite sandwich structure was the British aircraft De Havilland Mosquito that was produced during the Second World War. Sandwich made of veneer faces and balsa core was mass produced [2]. Some years later, in 1943, the American military aircraft Vultee BT-15 was developed with a sandwich fuselage made of glass fibre skins and balsa core [16]. Balsa was the first core material to be industrially used and is still in use in applications where the weight is not critical like in cruising yachts or rotor blades for wind turbine.

Research and development of sandwich structures started in the 1940's and continues nowadays with the aim to improve the damage behaviour and the weight efficiency creating new materials and concepts. Honeycomb cores were already developed in 1905 in Germany, but they were used as core material first at the end of the 1940's mainly in aerospace structures [17]. Honeycomb cores are produced with different forms and materials, for instance aluminium and aramid paper, and offer the best weight specific material properties. However, due to their high costs, their use is limited to aerospace applications. In the early 1940's, the first polyvinylchloride (PVC) was developed in Germany. But it took about 15 years to be used in industrial applications due to the softness of the first developed foams. Due to the improvement of manufacturing processes and material properties of the PVC and polyurethane (PU) core materials, they are today commonly used in many low and medium cost industrial applications. In addition to the further development of core and adhesive materials, the use of composite materials as skins has given an important push to the sandwich structure to be used in different sectors.

Nowadays composite sandwich structures find a broad range of applications in commercial aircraft and satellite structures. While almost all satellite structures are made of honeycomb sandwich constructions [16], the use of composite sandwich structure in commercial aviation is limited to secondary structures [2]. Due to the strict safety requirements, it is important to understand the structure behaviour after damage occurrence. Therefore, every structure should prove that every in-service damage occurrence doesn't lead to failure or excessive structural deformation until the damage detection. Due to the lack of material understanding and design methods as well as in-service detection methods of damages in sandwich structures, the aerospace industry is cautious to use the sandwich construction in primary aircraft structures. A variety of sandwich composite applications in commercial aircrafts can be found, for example radome, engine cowling, leading and trailing edges, control surfaces, floor panels and last but not least landing gear doors. Most of these components use Nomex honeycomb as core material, while ROHACELL® PMI-foams are only used as manufacturing aid like in the hat profiles for the Airbus A340 and A380 rear pressure bulkheads. Depending on the location of the sandwich component in the aircraft different design requirements are applied. External structures like the landing gear doors and engine cowling are exposed to runway debris. Choosing materials for sandwich components should also consider the electromagnetic transmissibility for radar and avionics systems, and the fire, smoke and toxicity requirements (FST). Figure 2 shows the use of sandwich composite materials in the turboprop aircraft ATR 72-500/-600. In the recent years, the use of sandwich composites in helicopters has known an increasing trend especially in the rotor blades and the airframes. For example, the rotor blades of the Eurocopter EC135 are made of GFRP sandwich, while the secondary structure and small parts of the primary structure are made of carbon and aramid fibre sandwich [18]. The honeycomb is also the most used core materials in helicopters sandwich structures.

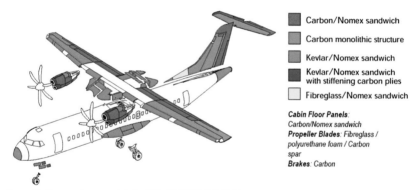

Figure 2. Composite materials in ATR 72-500/-600 [19]

The use of sandwich construction is well established in marine industry especially for high performance boats and leisure boats, in which foam core composite sandwich is extensively used to reach the weight requirements. One of the most known high performance boats is the monohull navy ship YS2000 Visby of the Swedish marine forces, which is made primarily of composite sandwich construction making advantage of the non-magnetic properties of the materials and the high energy absorption combined with low weight.

In transportation sector sandwich structures can be found in many applications, for example, the structure of special vehicles like ambulances or fire engines is mainly made of sandwich construction profiting from the weight advantage and the thermal and acoustic insulation properties of the foam core. In addition to road transportation, composite sandwich structures are widely used to build the front cabs of the locomotives of high speed trains like the French TGV and the Japanese Shinkansen 500 series [16].

The wind power energy sector is a good demonstrator for the massive use of the sandwich construction profiting from the high weight specific bending properties. Typically GFRP is mainly used for the skins combined with foam cores or balsa wood, CFRP is also used where more stiffness and strength is needed. The sandwich construction in wind turbine is mainly used in the rotor blades but it can also be used for the manufacturing of the nacelle and turbine housing. Depending on the manufacturer, different architectures of the rotor blade can be found. In Figure 3 the cross-section of a Gamesa corp. rotor blade is presented. The sandwich structure can be found in the webs of the load carrying box girder and the aerodynamic skins of the rotor blade, where PMI- and PVC- foams are used as core materials.

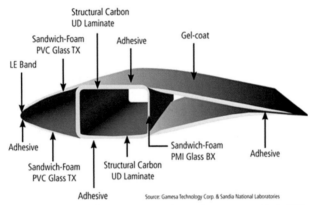

Figure 3. Schematic presentation of a cross-section of a wind energy rotor blade [20]

2.1.3 Materials and manufacturing

Sandwich constructions consist basically of at least three components; the faces, the core and the adhesive joints. The faces should carry the in-plane tensile and compression loads resulted from the bending loads. The core should carry shear loads and out-of-plane compression and maintains the distance between the skins constant. The adhesive joints bond the two other components and enable the load introduction between the core and the faces. Depending on the application and the requirements different materials can be combined to a sandwich construction. These materials diversity enables to adapt the sandwich properties on the target requirements and to reduce materials costs.

Face materials

The face sheet is thinner than the core and the material differs generally from the core material as it should have superior mechanical properties and fulfil some key requirements like high tensile- and compression strengths, high stiffness, good impact resistance, good surface finish and last but not least good environmental resistance (high temperature, chemicals etc.). In addition to these key requirements other criteria like material costs, processability and some environmental aspects play an important role in the choice of the facings material.

The typically used materials for the face sheets can be divided in metallic and non-metallic materials. Steel, aluminium and titanium are the most used metallic materials. They have high stiffness and strength as well as good impact behaviour, but due to the poor drape properties of metallic sheets, the processing is very complex and limited to flat or simply curved panels. Wood, paperboard for packaging, fibre reinforced and non-reinforced plastics are commonly used as face materials. Fibre reinforced plastics are the most used face sheet materials due to the high weight specific properties, the good processing and the flexibility to adapt the material properties to the loading conditions. If composite face sheets are used, the sandwich construction can be called composite sandwich structure, which is the case in this work.

The most used reinforcing fibres in sandwich faces are carbon, glass and aramid fibres with a high dominance of the market to the glass fibre due to the unbeatable price performance ratio. Aramid fibres are typically used were high crash energy should be absorbed. The high performance carbon fibres are mainly used in aerospace applications like the rudder and stabilizer of different airplanes (ATR72, Airbus A320 …). Thermosetting matrix systems like EP, PH and VE are commonly used, while thermoplastic resins like PEEK or PPS slowly gain importance especially in automotive sandwich applications. In aerospace application thermoset epoxy resins combined with carbon fibre are typically used in external structures due to their high performance and thermal stability. In cabin applications, the FST-requirements oblige the use of materials with lower mechanical performance so that glass fibre with phenol resin are generally used.

Core materials

In order to reach the target of sandwich construction, the core material should fulfil some requirements. A low density of the core is important to increase the structure bending stiffness with minor weight penalties. High shear modulus and shear strength are vital to withstand global deformations and high transverse loads. To maintain the regular distance between the two faces and to reduce out-of-plane compression damages, enough strength and stiffness perpendicular to the faces are required. Moreover, the acoustic and thermal insulation properties of core materials are generally considered during the design of sandwich structures [13].

The core materials can be classified depending on the support type of the face sheets: we can distinguish between homogeneous support of the skins e.g. foam core types and balsa, and non-homogeneous support of the skins, which is classified into punctual, unidirectional and bi-directional support [21]. Figure 4 gives an overview about the different types of core materials. The most used under the "homogeneous skin support" -type are balsa wood, open-cell and closed cell foams. The latter can be prepared and shaped easily and the manufacturing of sandwich panels is very effective, as the bonding surface is homogeneous and it is possible to use out-of-autoclave manufacturing processes. Closed-cell polymeric foams offers compared to other core materials in the same category better lightweight performance and more flexibility concerning manufacturing and complex shaped geometry but they still have less strength and stiffness than balsa wood, metallic foams or honeycomb cores. The most known polymeric foams are polyurethane (PUR), Polystyrene (PS), Polyvinylchloride (PVC) and Polymethacrylimide (PMI).

In the category "non-homogeneous support of the skins" -cores three different types of cores can be identified: under "punctual support"-cores we can find truss cores or pin-reinforced cores with removed foam, corrugated cores belong to the "unidirectional support" -cores and honeycomb cores belong to the "bi-directional support" -core type. Nomex and aluminium honeycomb are widely-used in technical applications especially in aerospace applications. Honeycomb cores offer the highest weight specific mechanical properties [13] but they have an orthotropic mechanical behaviour and the manufacturing is complex due to the open cell structure.

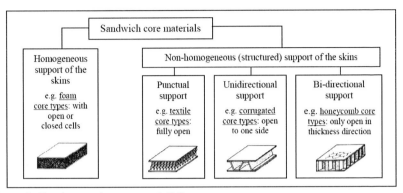

Figure 4. Classification of core materials [21]

Adhesive joint

To achieve the sandwich concept, an adhesive medium is essential to bond the skins to the core and to enable the load transfer between the different components. Choosing the right adhesive material depends mainly on the sandwich components, the manufacturing methods and the environment where the structure should be used. The adhesive should fulfil some requirements like high tensile and shear strengths, good fatigue behaviour, low-temperature sensitivity, low curing shrinkage, ductility and last but not least environmental durability [13]. For foam or balsa core, a low viscosity (to a defined limit) is needed to enable the adhesive to flow from rich area to dry area while bonding pressure application or resin injection or infusion.

Adhesive materials exist in different forms in the market namely paste, liquid, film and last but not least powder form. Depending on the requirements, a suited adhesive can be chosen. Polyurethane (PUR) adhesives are the most widely used adhesives for bonding with the co-bonding method (finished skins are simply bonded to the core). They exist as paste or liquid with different properties and offer good adhesion to a wide range of materials. Polyester and vinylester resin are widely used for sandwich with composite skins in applications outside the aerospace industry and they can be used with co-bonding or co-curing and co-infusion bonding (simultaneous curing of the adhesive layers with the composite skins) methods as the resins are commonly used for the face sheets. Epoxy resins are especially used in aerospace applications as adhesive medium due to the good mechanical properties and the high temperature resistance. Phenolic resins are used in sandwich structures in the cabin with some additive to fulfil the FST requirements.

Manufacturing

Choosing a suited manufacturing process for sandwich structures depends mainly on the used materials. We can distinguish between co-bonding, co-curing and co-infusion processes. Under co-bonding manufacturing of sandwich structures we can understand a process that applies temperature and pressure to activate the adhesive material and to bond the sandwich components. This process can be applied to sandwiches with metallic skins or cured composite face sheets. The adhesive layer is placed between the core and the skins, and then the bonding is activated by increasing the temperature and the pressure, whether with a vacuum setup or a press. Finally, the sandwich panel is cooled down and demoulded. A co-curing process is a manufacturing process for composite sandwich structures at which the sandwich components are bonded during the curing of the face sheets. A wet hand lay-up or prepreg lay-up setups can be used, where the surplus of resin will bond the face sheets and the core. A pressure and a vacuum are generally applied to improve the compacting and to remove the excess resin. This process is suited for solid cores and open-hole cores like honeycomb. However, special care should be taken when manufacturing honeycomb sandwich, as there is a risk of face sheet deformation (out-of-plane waviness) on the open side of the mould due to missing support of the uncured skins.

Co-infusion processes include all types of liquid composite moulding (LCM) processes like resin transfer moulding (RTM) or vacuum assisted resin infusion (VARI). They are well suited for foam core sandwich composites as complex structures can be produced in one shot and good quality can be reached. Moreover, using dry fibre preforms and LCM reduces process time and material costs compared to the prepreg technology [22]. However, LCM processes are only conditionally suited for honeycomb sandwich as the honeycomb open cells should be pre-sealed to avoid filling the honeycomb structure with liquid resin. Figure 5 shows a schematic setup of a VARI-process for foam core sandwich panels. Closed cell foam cores are well suited for this process as the continuous support of the textile skins overcomes the drawbacks of the honeycomb cores. This process is widely used in marine and wind turbine industry as it enables to manufacture large and complex structures. The preform consisting of core and dry fibre layers is placed into a mould. Inserts and other options like monolithic profiles, ramps or transitions can be also integrated. Special binders are typically used to fix the different components during preforming. After placing the different manufacturing aids, the setup is sealed and a vacuum is applied. The resin valve is opened and the face sheets are impregnated. After full impregnation, the curing starts following the curing cycle provided by the resin manufacturer. Finally, the cured part can be demoulded. The VARI-Process needs only a half-open mould, so that the surface with high quality requirement should lie on the mould side. The RTM-process needs, however, a two parts mould that creates the compacting pressure. The resin can be injected under vacuum (VARTM) or at a defined pressure.

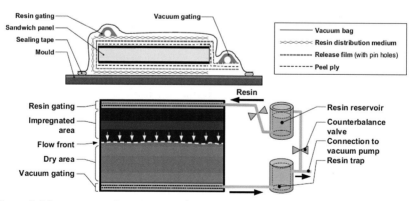

Figure 5. Schematic setup of a VARI-process for sandwich structures [22]

2.1.4   Fundamentals of sandwich theory

The sandwich principle is based on the Huygens-Steiner theorem. By increasing the distance between the face sheets and the bending axis of the sandwich, the area moment of inertia and therefore the bending stiffness of the structure are dramatically increased since the area moment of inertia increases with the square of the face sheet distance to the sandwich neutral axis as shown in the following equation for the calculation of the bending stiffness of a beam.

$$D = EI = \int_{-\frac{b}{2}}^{\frac{b}{2}} \int_{-\frac{h}{2}}^{\frac{h}{2}} E z^2 dz dy \, . \tag{2.1}$$

Where E and I are the elastic modulus and the area moment of inertia of the beam; z and y are the integration variables, and h and b are the height and the width of the beam respectively. This equation can be applied to calculate the bending stiffness of sandwich structures. The sandwich effect is only possible when the core and the facings are bonded together. This requirement is not fulfilled anymore and the sandwich structure loses its stiffness if the adhesive fails. For a symmetrical sandwich cross-section, which is the case when the skins have the same thickness and material properties, the bending stiffness can be calculated using equation 2.2, where b is assumed to be equal to one. The used sandwich parameters are illustrated in Figure 6.

$$D = \int E z^2 dz = \frac{E_f h_f^3}{6} + \frac{E_f h_f h_d^2}{2} + \frac{E_c h_c^3}{12} = 2D_f + D_0 + D_c. \tag{2.2}$$

Where $2D_f$ and D_c are the bending stiffness of the face sheets and the core about their individual neutral axes and D_0 is the bending stiffness of the face sheets about the neutral axis of the sandwich. E_f and E_c are the elastic moduli of the face sheets and the core, h_f and h_c are their thicknesses and h_d is the distance between the centroids of the faces.

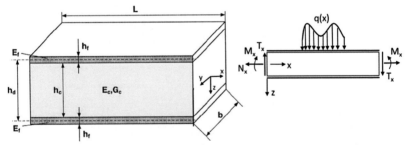

Figure 6. Sandwich parameters for bending stiffness calculation and sign convention for stresses (left), forces and bending moments (right) [7]

According to Zenkert [13] equation 2.2 can be simplified if the sandwich structure consists of a thin face sheet and a weak core so that the bending stiffness of the sandwich beam is reduced to D_0. This is only possible if the following assumptions are satisfied:

- Thin face sheet assumption:

$$\frac{2D_f}{D_0} < 0.01, \; if \; 3\left(\frac{h_d}{h_f}\right)^2 > 100, \; or \; \frac{h_d}{h_f} > 5.77. \tag{2.3}$$

- Weak core assumption:

$$\frac{D_c}{D_0} < 0.01, \; if \; \frac{6E_f h_f h_d^2}{E_c h_c^3} > 100. \tag{2.4}$$

The bending stiffness of sandwich beam can now approximately be written as:

$$D = D_0 = \frac{E_f h_f h_d^2}{2}. \tag{2.5}$$

Equation 2.5 shows that the distance of the face sheets to the neutral axis is the main influencing factor of the bending stiffness of sandwich beams. The dependence of the stiffness on the distance between the face sheets is known as sandwich effect. Dan Zenkert demonstrates in his book [13] the sandwich effect by comparing the bending strength and stiffness of a monolithic beam with sandwich beams, where the thickness of the face sheets is similar to the thickness of the monolithic beam and the core thickness was increased. Doubling the core thickness leads to the doubling of the sandwich bending strength and quadruplicating of the bending stiffness. It is clear that a sandwich construction increases the stiffness of a shell structure dramatically and thus improves its buckling strength. The sandwich effect is depicted in Figure 7.

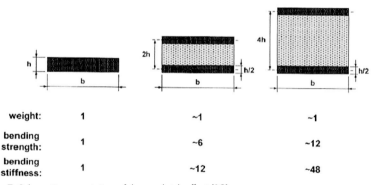

Figure 7. Schematic presentation of the sandwich effect [13]

A sandwich structure works similar to an I-beam. The face sheets, similar to flanges, carry mainly the normal stresses due to bending and the core, analogue to the web, carries primarily the shear stresses. The normal stresses in the skins according to sign convention in Figure 6 are

$$\varepsilon_f = \frac{M_x z}{D}, \; and \; thus \; \sigma_f(z) = \frac{M_x z E_f}{D}. \tag{2.6}$$

This equation gives a compression loaded and a tensile loaded face sheet.

The normal stresses in the core are calculated with the following equation:

$$\varepsilon_c = \frac{M_x z}{D}, \text{ and thus } \sigma_c(z) = \frac{M_x z E_c}{D}. \tag{2.7}$$

The stresses vary linearly in each material but with different slopes due to the stiffness gap between the materials, which creates a jump at the transition between the two different components.

The stresses and strains due to an in-plane load N_x are then:

$$\varepsilon_{x0} = \frac{N_x}{E_{f1}h_{xf1} + E_{f2}h_{xf2} + E_c h_c}, \tag{2.8}$$

and thus $\sigma_c = \varepsilon_{x0} E_c$, $\sigma_{f1} = \varepsilon_{x0} E_{f1}$ and $\sigma_{f2} = \varepsilon_{x0} E_{f2}$.

With ε_{x0} is the normal strain at the neutral axis of the sandwich beam. Stresses and strains due to bending and in-plane loads can be superposed. The distribution of in-plane stresses in the sandwich components are illustrated in Figure 8 for different levels of approximations.

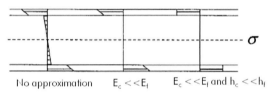

No approximation $E_c \ll E_f$ $E_c \ll E_f$ and $h_c \ll h_f$

Figure 8. In-plane stresses due to global bending [13]

As the shear stresses do not vary linearly along the sandwich cross-section Zenkert [13] proposed simplified general formulations to calculate the shear stress distributions in the face sheets and the core using the following equations:

$$\tau_f(z) = \frac{T_x}{D} \frac{E_f}{2} \left[\frac{h_c^2}{4} + h_c h_f + h_f^2 - z^2 \right]. \tag{2.9}$$

$$\tau_c(z) = \frac{T_x}{D} \left[\frac{E_f h_f h_d}{2} + \frac{E_c}{2} \left(\frac{h_c^2}{4} - z^2 \right) \right]. \tag{2.10}$$

Where $T_x = dM_x/dx$ is an equivalent transverse force describing the change of the bending moment in the x-direction of the sandwich beam. Two shear stress values are important for a sandwich structure. The first one is the maximum shear stress $\tau_{c,max}$, which is located in the sandwich neutral axis ($z = 0$). The second one is the maximum shear stress in the face/core interface ($z = \pm h_c/2$), which corresponds to the minimum shear stress in the core ($\tau_{c,min}$) and the maximum shear stress in the face ($\tau_{f,max}$).

$$\tau_{c,max}(z = 0) = \frac{T_x}{D} \left[\frac{E_f h_f h_d}{2} + \frac{E_c h_c^2}{8} \right], \text{ and} \tag{2.11}$$

$$\tau_{c,min} = \tau_{f,max} = \tau_c \left(z = \frac{h_c}{2} \right) = \frac{T_x}{D} \left[\frac{E_f h_f h_d}{2} \right]. \qquad (2.12)$$

The shear stress in the core is nearly constant along the core thickness (ratio between the minimum and maximum shear stress <1%) if the following condition is satisfied:

$$\frac{4 E_f h_f h_d}{E_c h_c^2} > 100. \qquad (2.13)$$

If the two main assumptions for simplification of sandwich analysis are applied to stress calculations, the previously outlined equations can be further simplified. For the assumption of a weak core material ($E_c \ll E_f$) the stresses are then [13]

$$\sigma_c(z) = 0, \qquad \sigma_f(z) = \frac{M_x z E_f}{(D_0 + 2D_f)}, \qquad (2.14)$$

$$\tau_c(z) = \frac{T_x E_f h_f h_d}{2(D_0 + 2D_f)} \quad \text{and} \quad \tau_f(z) = \frac{T_x E_f}{4(D_0 + 2D_f)} \left(\frac{h_c^2}{4} + h_c h_f + h_f^2 - z^2 \right). \qquad (2.15)$$

If the assumption of thin face sheets ($h_f \ll h_c$) is additionally considered the equations 2.14 and 2.15 are further simplified with [13]

$$\sigma_c(z) = 0, \qquad \sigma_f(z) = \pm \frac{M_x}{h_f h_d}, \qquad (2.16)$$

$$\tau_c(z) = \frac{T_x}{h_d} \quad \text{and} \quad \tau_f(z) = 0. \qquad (2.17)$$

The effect of these assumptions on the shear stress distribution is painted in Figure 9. From Figure 8 and Figure 9 it can be seen how the load carrying work is divided between the core and the face sheets.

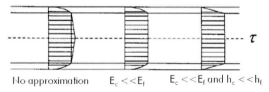

No approximation $E_c << E_f$ $E_c << E_f$ and $h_c << h_f$

Figure 9. Shear stresses due to global bending [13]

2.1.5 Failure modes of sandwich structures

Sandwich structures can experience different failure modes depending on the loading conditions, used materials and structure geometry. Stability failure, material damage as well as debonding are the most frequent damage types. They can be critical and limit the mechanical performance of the structure. For that reason they should be checked against all relevant load cases and analysed to determine the limits of the structure. The most common damage modes are summarised by Zenkert [13] and depicted in Figure 10.

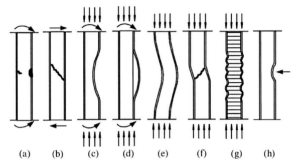

(a) (b) (c) (d) (e) (f) (g) (h)

Figure 10. Damage modes in sandwich beams and panels: (a) face yielding/fracture, (b) core shear failure, (c) and (d) face wrinkling, (e) general buckling, (f) Shear crimping, (g) face dimpling and (h) local indentation [13]

These damage modes are briefly explained with respect to the damage behaviour of composite sandwich structures.

(a) Yielding or fracture of the face sheet in tension or compression

Subjected to high bending loads composite face sheets exhibit typically compression or tension damage. Identifying face sheet yielding or fracture depends on the used damage criterion as various damage criteria for composite materials are available. Generally, the stresses in the different layers are calculated using equations 2.6 and 2.9 and then applied to the damage criterion. Alternatively, the maximum stresses in the face sheets can be compared to the ultimate stress in tension and compression. However, both face sheets should be verified since they are differently loaded. As most of the core materials have a higher yield and fracture strain than the face sheets, the face sheets would generally fail long ago before the damage limit of the core is reached, which makes the use of a failure criterion for the core under bending load unnecessary.

(b) Core shear failure

As previously mentioned the core carries most of the transverse force. The resulted shear stresses are superposed with the normal stresses. As fracture criterion the maximum core shear stress is compared to the ultimate stress of the core. Following Zenkert [13] the shear fracture criterion can be written

$$\bar{\tau}_{c,xz} = \left[\left(\frac{\sigma_{c,x}}{2}\right)^2 + \tau_{c,xz}^2\right] < \hat{\tau}_{c,x} \quad \text{and} \quad \bar{\tau}_{c,yz} = \left[\left(\frac{\sigma_{c,y}}{2}\right)^2 + \tau_{c,yz}^2\right] < \hat{\tau}_{c,y}. \tag{2.18}$$

Where $\hat{\tau}_{c,x}$ and $\hat{\tau}_{c,y}$ are the allowable shear stresses of the core in x and y direction respectively. If the weak core assumption ($E_c \ll E_f$) is used the direct stress is neglected and the shear stress can be calculated with equation 2.17. The shear stress induces a tensile stress equal to the shear strength of the core at a 45° angle from the x-direction and leads to a 45° inclined cracks. These cracks are characteristic for pure shear failure and are called shear cracks [13].

(c and d) Face wrinkling

Face sheet wrinkling can occur in practical cases when a sandwich panel or beam is subjected to compressive buckling or if the face sheet is affected by in-plane compression due to bending loads. This kind of damage can appear in two forms: outward wrinkling if the tensile strengths of the adhesive joints or the core are lower than the compressive strength of the core, inward wrinkling if the compressive strength of the core is lower than the tensile strength of the core and the adhesive joint. Zenkert proposed the following formula to calculate the critical wrinkling stress for sandwich beams

$$\hat{\sigma}_{f,FW} = \frac{1}{2}\sqrt[3]{E_f E_c G_c}. \tag{2.19}$$

(e) General buckling

General buckling is a stability damage mode that may occur when the structure is subjected to in-plane compression. Buckling of structures should be avoided as it leads to the loss of the structure functionality. The critical buckling load can be the ultimate design load as a buckled structure may not withstand any further loading. If the deformation is controlled and the load continues to deform the structure after buckling and load drop, the structure would fail by compression failure of the face sheet side under compression, or by face sheet wrinkling on the compressed side, or by core shear failure.

Impact related damages like debonding, crushing or shear cracks may degrade the structure stiffness and have a detrimental impact on the sandwich stability. Local damages like core crushing, debonding and face sheet failure may intensify local instability problems, while global damages like shear cracks can degrade the global buckling strength. For these reasons it is important to investigate the compression after impact (CAI) behaviour of sandwich structures exposed to impact loading.

(f) Shear crimping

Shear crimping failure is a form of general buckling and occurs in sandwich with a core having low shear modulus. This failure mode is more likely to occur in post-buckle stage due to large out-of-plane deformation and induced transverse forces. This damage is marked by core shear failure often followed by immediate face sheet damage. The shear crimping is a sort of a short wavelength form of antisymmetric wrinkling. The critical shear crimping load for a sandwich beam under in-plane compression can be calculated using the following equation:

$$P_{SC} = b h_c G_c. \tag{2.20}$$

With b is the width of the sandwich beam.

(g) Face dimpling

Face dimpling or inter cellular buckling is local instability phenomena that may occur in sandwich structures with discontinues face sheet support like honeycomb or corrugated cores. The face sheet without core support buckles locally while the supporting materials doesn't experience any damage. Formulae to calculate critical dimpling load for different core geometries are given in [13].

(h) Core indentation

Core indentation damage may occur when concentrated out-of-plane load are introduced into a sandwich structure over insufficiently large area, such as e.g. fitting, joints or due to impact loading. When point loads are applied, the face sheet acts like a plate on elastic foundation. Once the compressive strength of the core is exceeded, the core will fail. To avoid this damage, a minimum sufficient area for load introduction A_L should be estimated using the following equation [23]:

$$A_L = \frac{F_L}{\hat{\sigma}_c}. \qquad (2.21)$$

Where $\hat{\sigma}_c$ is the compressive strength of the core in loading direction and F_L is the punctual load. This formula is not applied to impact loading as impact is rather a point load than surface load and the impact load is a superposition of global bending load and local indentation load. Further discussion of impact behaviour is available in section 2.3.

2.2 Through-the-thickness reinforcement of foam core sandwich structures

2.2.1 Motivation and examples from the literature

The main advantage of using sandwich structures is the huge increase in flexural stiffness with very moderate weight penalty, which outperforms the performance of other monolithic composite constructions. Honeycomb sandwich composites have been established as state-of-the-art core material for aerospace applications due to the high weight-specific properties. However, honeycomb cores exhibit some drawbacks that limit their use to non-primary aircraft structures and to small aircrafts with fewer restrictions. Honeycomb sandwich structures are susceptible to moisture take-up in the open-cell structure, which can lead to corrosion and other invisible damages like debonding and buckling. Moreover, the edgewise support of composite skins and the anisotropic properties of the core make it prone to impact loading and to local face sheet instability known as face wrinkling.

Considering the manufacturing of honeycomb sandwich, the hole structure of honeycomb provides a bi-dimensional support to the face sheet, which can result in uneven surface, especially when thin skins are bonded to a core with large cells using a co-curing process. Moreover, manufacturing of honeycomb sandwich is quite expensive and inefficient, as low-cost out-of-autoclave manufacturing may not be possible. In addition, complex shaped structures or pre-set curvatures may be hard to achieve due to Poisson effects [13].

In the last years, closed cell foams like PMI foam materials have been experiencing a dramatic improvement regarding the mechanical and thermal properties. Although honeycomb cores outperform PMI foams in terms of stiffness and strength-to-weight ratios, they still have many advantages that make them attractive [13]. They have a heat resistance up to 210 °C [24], which enables the manufacturing with epoxy systems (prepreg or VARI) in a 180 °C environment. The fine closed cell structure overcomes the issue of moisture take-up, provides an even and smooth bonding surface and simplifies the surface preparation and shaping. Despite the high material costs of PMI foams, the easy processing and manufacturing of foam core sandwich composite by using liquid resin injection methods make the total costs (materials + manufacturing) 25% lower compared to a similar structure made of honeycomb and epoxy prepreg [25]. Moreover, PMI foams exhibit similar impact damage resistance and visual inspection detectability as Nomex honeycomb [25].

Despite the good weight-specific mechanical properties of honeycomb structures and the recent improvements in performance of foam cores, sandwich structures are still prone to out-of-plane loading, for instance low-velocity impact and indentation, and have serious issues with debonding and structure integrity. In recent years, many solutions have been proposed to overcome these issues. One of these solutions is combining the properties of the honeycomb and the foam cores by filling the cavities in the honeycomb with expandable foam material. This improves the performance of the core by increasing the buckling stability of the cell walls of the honeycomb, creating a homogenous support to the skin and enabling the use of out-of-autoclave manufacturing processes. The out-of-plane properties and the debonding resistance are highly enhanced [26, 27] but they are coupled with unavoidable weight penalties [28]. A similar concept to improve the mechanical performance of core materials is the 3D|CORE™. 3D|CORE™ is a foam having hexagonal channels that are filled with resin during infusion or lamination leading to the creation of hard resin structure in the foam after curing of the sandwich panel. This concept improves the rigidity and the strength of the structure dramatically and fastens the manufacturing process due to resin channels, when vacuum assisted resin infusion is used. Nevertheless, the weight is highly increased and the impact performance is degraded compared to a standard foam core sandwich [29]. Figure 11 shows samples of foam-filled core and 3D|CORE™.

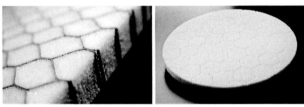

Figure 11. Left: foam filled honeycomb core [30]; right: 3D|CORE™ [31]

In order to enhance the mechanical properties of foam core sandwich composite without increasing the foam density or leading to significant weight penalties, the through-the-thickness reinforcement of foam core has demonstrated to be an effective way to reach this goal. Some of these methods are demonstrated in Figure 12. By placing composite webs in the core like TT-Profiles or the foam-filled corrugated sandwich the stiffness of the sandwich panel is increased and the core cracks could be stopped leading to improved structure integrity. To reinforce the core-skin interface and improve the overall strength and stiffness while maintaining the weight penalties under control, the through-the-thickness reinforcement by means of stitching, tufting or Z-pinning has shown very promising results [4]. One of these techniques, the X-cor™ sandwich, has been already used in the tail cone of the helicopter UH60M Black Hawk.

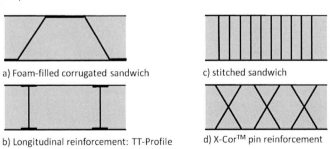

a) Foam-filled corrugated sandwich

c) stitched sandwich

b) Longitudinal reinforcement: TT-Profile

d) X-Cor™ pin reinforcement

Figure 12. Examples of foam core reinforcements

19

Many technologies to insert pin reinforcements into the foam core have been developed based on industrial traditional methodologies. Stitching, tufting and Z-pinning processes (Figure 13) belong to the most used pinning technologies. The used Tufting and Stitching techniques are based on standard textile processes. In both processes dry fibre threads are inserted into the sandwich connecting the core and the face sheets to each other, which increases the delamination and debonding resistance of the sandwich [32, 33]. The dry fibre reinforcements can be inserted automatically with the possibility to change the insertion angle and reinforcement density. While Stitching needs a two-sided access to form a loop or an interlocking pattern with another thread on the opposite side of the panel, the tufting process requires only a one-side access. Only a special elastic foam tooling is needed to keep the reinforcing thread in the sandwich preform after insertion, which enables the needle to withdraw. These processes are suited for out-of-autoclave manufacturing techniques in which the dry fibre reinforcements are impregnated and co-cured.

Another pinning methodology is the insertion of cured rods in the through-the-thickness direction of the sandwich panel. This step takes place prior to the manufacturing of the sandwich panel. The rods can be inserted whether into the core or through the skins and the foam core. Pins made of different materials can be inserted, ranging from pultruded carbon-epoxy resin rods to titanium alloy pins. The ultrasonically assisted Z-Fiber® pinning process developed by Aztex Inc. is one of the leading technologies in this field. It is nearly the same process used to insert pins in uncured laminates. The Z-pins are placed with a defined pattern (insertion angle, spacing...) inside a foam carrier to provide lateral support. Afterwards the pin carrier is placed over the sandwich preform, and an ultrasonic insertion gun generating high frequency compressive waves pushes the pins progressively into the sandwich preform leading to the collapse of the foam carrier. When all pins are inserted, the foam carrier is removed and the pin extensions are shaved off.

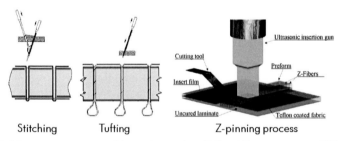

Stitching Tufting   Z-pinning process

Figure 13. Schematic presentation of the stitching, tufting [34] and Z-pinning process [35] applied to sandwich structure

Regarding the mechanical properties, adding pins in thickness direction of foam core sandwich structures leads generally to the same trends independent of the used reinforcing technology. Most of the available reports in open literature promises improvement of the mechanical performance of foam core sandwich structure, when through-the-thickness reinforcements are used. Differences can be observed in the damage modes and the resulted mechanical properties, which depend on the pinning methods and the used materials. The overall tendencies are briefly presented in this section. A detailed discussion can be found in the following chapters.

Studying the compression performance of pin-reinforced sandwich structures [36, 12, 37] showed an increase in compressive strength and stiffness with the increase of the pin volume fraction in the core. The inclination angle of the pins has a major influence on the compression performance as it changes the load distribution in the core. Orthogonal pins lead to the highest compression properties, while inclined pins are favourable for the shear properties. Blok et al. [38] investigated the edgewise compression performance of foam core sandwich structures tufted with aramid fibre. They concluded that the tufting yarns stabilize the fracture occurrence compared to untufted specimens, which increases the debonding resistance and the specific energy absorption. Similar results were also reported by Henao et al. [34]. The improvement of the compressive properties is generally accompanied by the enhancement of the tensile [36, 39], bending [34, 40] and shear properties [36, 41] as well as the resistance to peeling [39, 32]. However, the change in mechanical properties depends strongly on the inclination angle [42], the volume fraction of the reinforcements and the pin pattern so that an optimal configuration should be found for every loading condition. In this contest Lascoup et al. [12] and Du et al. [41] investigated the effect of loading direction on the shear properties. It was found that the reinforcements work efficiently when the pins are aligned opposite to the shear loading direction. Regarding the impact performance many works registered an improvement of the impact resistance by the integration of stitches or pins into the foam core. Baral et al. [43] reported that a pin-reinforced foam core sandwich structure under water slamming loads outperformed a honeycomb core sandwich panel with the same areal weight and skin materials. However, improvement of the mechanical properties by using through-the-thickness foam reinforcements is not always evident. Nanayakkara et al. [44] investigated the impact performance of z-pinned sandwich. They registered only a minor improvement of the impact performance and post-impact compression properties at high impact energies and no improvement at low impact energies. The observed behaviour could lie on the used orthogonal pins as the pin insertion angle is not favourable for impact loading. Similar behaviour was observed by Kim et al. regarding the fatigue behaviour under bending load. While the use of stitched sandwich increased the sandwich strength and stiffness under static bending load compared to non-stitched reference sandwich, the material degradation of the stitched specimens in fatigue was higher than the degradation measured in reference specimens for the same fatigue life.

2.2.2   Tied foam core technology

In this work the Tied Foam Core (TFC) pinning technology developed by Airbus Group has been used to insert pins into the foam core. The TFC-technology is a pinning process that automatically produces pin-reinforced foam cores prior to the manufacturing of the sandwich panel and it uses dry fibre reinforcements instead of pre-cured rods, which increases the automation level. Compared to other industrial pinning processes the TFC-technology increases the manufacturing efficiency, as no pre-cured pins are needed, enables the use of a broad range of fibre and foam materials and offers a high geometrical flexibility, since pinning of core panels with curvatures, chamfers or cut outs is possible. As stitching of the foam is done by a needle, the pinning of thick foam cores with high density is now possible. Moreover, using dry fibre pins make the manufacturing of pin-reinforced foam core sandwich structures using low-cost out-of-autoclave processes, like resin transfers moulding (RTM) or vacuum assisted resin infusion (VARI), more attractive, since the pins are impregnated and co-cured during the manufacturing of the sandwich panel. However, it is also possible to use prepreg skins and autoclave cure. In this case, the inserted pins should be impregnated and cured separately before applying the prepreg layers. In addition, the skins are not stitched by the needle, which avoids face sheet damaging. Figure 14 shows two examples of pin-reinforced foam cores manufactured with the TFC-technology. The pin extensions of the glass and carbon fibre pins can be identified in the figure. When vacuum is applied during the VARI-process the dry fibre extensions are compressed and fixed in the interface between the core surface and the skin.

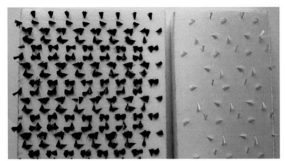

Figure 14. Examples of pin-reinforced foam cores. Left: carbon fibre, right: glass fibre

The TFC-pinning process consists of four main steps [45]: a needle stitches through the foam and hooks in the roving, which is already on the other side of the foam plate. Then the needle is pushed back and the fibre bundle is cut with predefined extensions. The different steps are shown in Figure 15. The stitch angle, pin-density, pin-diameter and pin-pattern are variable and can be changed depending on the required mechanical properties. The process enables to have straightened fibre, which improves the mechanical performance of the pins. After manufacturing of the sandwich panel, the roving extensions lie in the interface between the face skins and the foam core connecting the upper and lower skins to each other and creating an additional load path. This would enhance the fracture toughness of the interface [46] and slow down the crack propagation in the interface through fibre bridging and crack redirection.

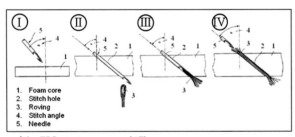

Figure 15. Steps of the TFC-pinning process [47]

Figure 16 shows a cross section of foam core sandwich specimen reinforced with pins made of glass fibre. The folded pin extension that lies between the skin and the core surface is highlighted in the figure. The pinning process has few negligible drawbacks on the panel quality. After curing of the sandwich panel, the pin extensions could leave minor out-of-plane waviness in the face sheet on the vacuum bag side. Moreover, during stitching the needle damages the cell walls of the foam core. The damaged cells are then filled with epoxy resin during the infusion, which increases the pin diameter, creates a resin coat around the pin and limits its fibre volume content. The resin rich area can be clearly identified in Figure 16.

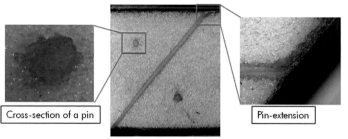

| Cross-section of a pin | | Pin-extension |

Figure 16. Cross section of a foam core sandwich reinforced with glass fibre pins after curing

2.3   Impact behaviour of foam core sandwich structures

This section deals with the low-velocity impact behaviour of foam core sandwich structures. The main damage modes are presented and explained. An insight into the most known analytical models to predict the low-velocity impact behaviour of sandwich structures is given, where the classification of the impact response types is in detail discussed. Finally, criteria for damage onset in foam core sandwich are introduced.

Many researches have been done to investigate the response of sandwich structures to low-velocity impact loading and to predict the damage modes that occur. Most of these investigations are summarised in the reviews provided by Abrate [48] and by Chai and Zhu [49]. Madenci and Anderson [50] compared the damage behaviour of honeycomb core and foam core sandwich specimens. They found out that, the investigated honeycomb sandwich configurations have better damage resistance than the configurations with foam core, and that the resin pools in the honeycomb sandwich specimen reduced the level of damage. Zhou et al. [51] investigated the perforation resistance of foam core sandwich. They concluded that perforation resistance of the tested specimens is very dependent on the properties of the foam core, namely the density and the cell structure, and a marine environment is detrimental for the perforation resistance. In another investigation done by Leijten et al. [52] the impact and CAI-behaviour of foam core sandwich panels representative for primary aircraft structures were studied. They concluded that core damage has no significant effect on the residual compressive strength except in case of shear cracks, which lead to more structure degradation.

In order to determine a suitable material combination of sandwich structure and to speed up the design process, many analytical models have been developed to predict the low-velocity impact response of sandwich panels and the failure occurrence [53-56]. These analytical models are suited to be used in preliminary design stages, as they help the engineers to define the initial sandwich configuration and to choose suitable materials. When accurate results concerning the impact response of sandwich structures and stress distribution in load introduction areas are needed, these methods reach their limits, as they consider simplified geometries and assumptions for boundary conditions that differ generally from the boundary conditions in a real sandwich structure. To overcome these issues the finite element method is used to predict the damage propagation and investigate the effects of boundary conditions on the impact response of the structure.

2.3.1 Impact failure modes

Damage tolerance is the main design requirement for load bearing aerospace structures. It describes the ability of structure to fulfil its function for a given period of time without catastrophic failure in spite of the pre-existence of a defined damage. For every structure allowable damage types and sizes are well defined, so that repair costs are reduced and maintenance periods are extended compared to a safe-life design methodology. If damage is beyond the allowable, it should be eliminated immediately. In order to design a damage tolerant structure, an extensive study of the damage scenarios during operations should be performed and it should be guessed how the structure would react to these damages. Some defects can grow under operating conditions and deteriorate the mechanical performance of the structure. These damages should be detected during routine inspections and repaired before reaching critical sizes. This behaviour is well-known as fatigue, which depends on the structure and the used material. For monolithic composites and composite sandwich structures foreign object impacts that occur during operations or maintenance were found to be the most critical as it could lead to larger invisible damage surface than from the outside visible damage and the structure shows brittle damage behaviour and nearly no plastic deformation unlike metallic materials.

In the aerospace we differ between ballistic, low-velocity and high-velocity impacts. Whereby it should be noted that low-velocity impacts are considered to be more critical, as they lead often to large hidden damages while the outside damage is too small or invisible. The ballistic impact creates visible perforation, which can be detected during visual inspections and its effect is limited compared to the consequences of large core damages. The damage resistance of composite sandwich structures is typically assessed by performing impact tests and evaluating the damage modes, while the compression after impact tests are performed to evaluate the damage tolerance of the structure. Minimum residual compression strength at defined damages is required for damage tolerant structures. In this section only the damage modes at low-velocity impact are discussed.

Low-velocity Impacts on foam core sandwich structures could create invisible damages like face sheet delamination, interface debonding, core crushing and core cracks. These damages could degrade the residual strength and the bending stiffness of the structure [52], so that they should be considered during the structure development phase. The impact load of a sandwich structure is characterised by the interaction between two main components, a local and a global load components. The sandwich response is a superposition of local indentation and global bending responses as shown in Figure 17. The overall impact behaviour depends strongly on the panel geometry, the boundary conditions, the mass and velocity of the impactor as well as the mass of the sandwich panel.

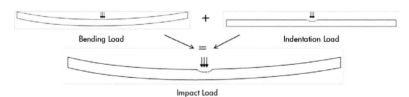

Figure 17. Response of sandwich structures under low-velocity impact: interaction between indentation and bending loads

The first load component is a localised compressive load that leads to face sheet failure, when a specific impactor force is reached. If the face sheet rupture load is not reached, core crushing and face sheet delamination below the point of impact occur, as only small impact energy is required to initiate these damages. The sandwich response is similar to the indentation response. The second load component is a global bending load compo- nent, which could lead, depending on the face sheet core thickness ratio, to oval shear cracks in the core. The shear crack damage occurs generally at a load lower than the face sheet rupture load. That means, if the face sheet is penetrated first, it is not possible that shear cracks initiate and propagate in the core [7]. The damage occurrence in impact loaded sandwich structures depends mainly on the impact energy and the core face sheet thickness ratio. At low impact energy local core crushing and face sheet delamination are likely to occur, if the face sheet rupture load is not reached. If the core face sheet ratio is low, typical for structure with thin core and thick skins, the bending load is dominating and damage is likely to occur in the core due to shear loading. These cracks in the core con- tinue to propagate in the lower interface and lead to debonding of the lower face sheet. This failure mode is critical, since core cracks are invisible during daily inspections and could lead to the loss of structure integrity. At high impact energies, skin penetration oc- curs in addition to the shear crack. This failure mode combines the global and local failure modes. If the core face sheet ratio of the sandwich structure is high enough, typical for thick core sandwich with thin face sheets, the compressive load will be dominating and only face sheet rupture with delamination, debonding of the top skin and core crushing below the impactor will be observed. This local failure mode is less critical for the integrity of the structure, as the failure surface is limited compared to the failure surface of the global failure mode and the failure is visible during standard inspections. A summary of the failure modes of impact loaded sandwich structure is shown in Figure 18. The Energy and the core face sheet ratio at transition between the different modes depends on the core and face sheet material combination. These failure modes are valid to simply sup- ported foam core sandwich panels.

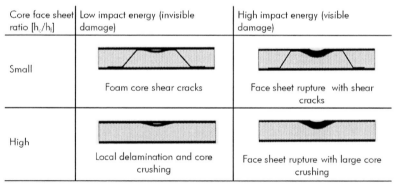

Core face sheet ratio [h_c/h_f]	Low impact energy (invisible damage)	High impact energy (visible damage)
Small	Foam core shear cracks	Face sheet rupture with shear cracks
High	Local delamination and core crushing	Face sheet rupture with large core crushing

Figure 18. Schematic presentation of damage modes after low-velocity impact (modified from [7])

2.3.2 Analytical analysis of the impact response of foam core sandwich structures

Impact behaviour is one of the major concerns of sandwich structures, which should be considered in early design stages. Costly and Extensive impact, fatigue as well as CAI-tests on coupon specimens are usually used to assess the impact resistance and the damage tolerance of composite structures. Dynamic FE-analysis with progressive damage evolution are very efficient to simulate the impact behaviour of sandwich structures, but they are very time consuming especially at preliminary design stages when many design parameters should be studied to find a suited sandwich configuration. However, using simplified analytical models to predict the impact behaviour of sandwich structures seems to be a successful way to reduce the design variations at early product development stages and to screen critical impact cases among different impact threats.

The classification of impact type has often been made by using the impact energy as reference value. But this assumption is only valid for impacts of similar conditions, this means similar structures with similar impactor velocity or mass. If the mass or velocity change dramatically while maintaining the impact energy constant, or if the panel materials are different, this kind of comparison is not any more valid, due to the increase of the inertia effects of the structure with the increase of the impact velocity. The standard classification of the impact response of a structure divides it into three different types: low-velocity, high-velocity and ballistic impact. This classification implies the use of impactor velocity to determine the type of the impact, but it neglects inertia effects and the material properties of the impact partners. Generally, an impact introduces an elastic wave that propagates in the structure. The wave propagation depends on the material damping, boundary conditions and the impact type. The ballistic impact has very short contact times, is characterised by a response dominated by three-dimensional dilatational waves and leads to visible perforation damages. High-velocity impact is characterised by a response dominated by flexural waves and shear waves and short impact times. The damage for this impact type can be compared to hail impact and runway debris damage. When the waves take a very long time to reach the boundary of the panel, the impact is predominated by the lowest vibration mode of the impactor-panel system. The impact response is quasi-static, since load-displacement relationship is similar to a static case. This impact type can be demonstrated by heavy tool drop scenario. The three different impact types are depicted in Figure 19. For a better classification of impact types, it is recommended to differentiate between ballistic and non-ballistic impact and to use the contact time as criteria. If the contact time is in order of magnitude of the time needed for a through-the-thickness stress wave to reach the other side of the plate, the impact is ballistic and is dominated by three-dimensional wave propagation. If the contact time is much larger than the time of a stress wave to travel the panel thickness, the impact is considered non-ballistic. For foam core composite sandwich the response of the core has a much earlier ballistic nature than the composite, that´s why the core properties are generally used to calculate the critical contact time.

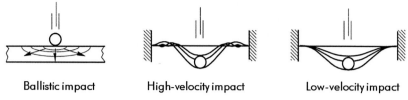

| Ballistic impact | High-velocity impact | Low-velocity impact |

Figure 19. Response types of plates under impact [57]

While it is common in the literature to classify the impact response based on the impact velocity, Olsson [57] demonstrated in his work that the impact response on composite plates for non ballistic impacts is governed by impactor vs. plate mass ratio and he derived new mass criteria for the two limiting impact cases. The small-mass impact response is a local response controlled by flexural waves. The large-mass impact response shows a global quasi-static behaviour. The impact response types based on the mass criteria are shown in Figure 20. The same mass criteria are applied to sandwich structures. Small mass impact takes place when the mass of affected plate area without boundary interactions is at least five times larger than the mass of the impactor. It should be noted that for orthotropic plates the first boundary hit by the wave is not always the closest to the impact location, and it is assumed that the affected area is elliptic. A quasi-static impact response is expected when the impactor mass is large than twice the total mass of the plate.

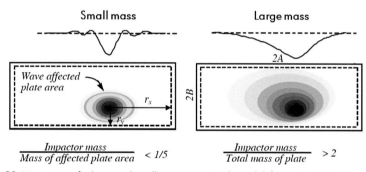

Small mass

Large mass

$$\frac{Impactor\ mass}{Mass\ of\ affected\ plate\ area} < 1/5 \qquad \frac{Impactor\ mass}{Total\ mass\ of\ plate} > 2$$

Figure 20. Mass criteria for large and small mass impact and panel deformation [55]

Different analytical models for impact damage prediction have been proposed in the open literature [58-60]. Commonly used are the spring-mass model, which uses springs to model the effective structural stiffness of the impactor-panel system, the energy-balance model that uses the conservation of the total energy in the system to find the maximum impact load and the modal superposition method. Chai and Zhu summarised in their review [49] the most known impact models for sandwich structures depending on the impact response type. The different impact types and the associated solution methods are presented in Figure 21. Models for small mass impact response only have to consider the indentation and local bending and shear stiffness of the panel, since the global panel bending is too small to consider the membrane stretching of the panel. Moreover, the impact behaviour is independent of the boundary conditions so that infinite boundary can be assumed. However, models for large mass impact response should consider the indentation, the global bending and shear stiffness of the panel, as well as the boundary conditions and the geometry of the panel. Local indentation and global bending can be estimated separately and then they can be superimposed [55]. Spring-mass models and energy based models are only used to estimate boundary controlled impact response (large mass). While the spring-mass models can predict the load-time history, the energy balance models, which utilises the principle of conservation of total energy of the panel-impactor system, can only estimate the maximum impact force. However, the accuracy of both models is deteriorated after damage onset. Foo et al. [60] coupled the energy-balance model with the law of conservation of impulse-momentum to take into account damage onset and extend the validity of the model. Olsson [55] provided an improved solution that considers local core crushing, delamination and large face sheet deflection. The modal superposition method is usually used for intermediate and small impactor plate mass ratio (small mass impact). It considers the impacted structure as a continuous infinite degree of freedom system. Olsson [61] proposed closed form solutions for monolithic plates and sandwich panels. He combined the solutions with a quasi-static delamination threshold load criterion and found a good agreement with experimental results.

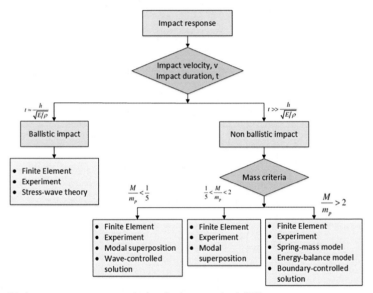

Figure 21. Impact response types and related solution methods [49]

In addition to the estimation of the load-displacement relationship of the impact behaviour, it is important to predict the critical loads for damage initiation to assess the impact resistance of the sandwich structure. If a critical load is smaller than the estimated maximum impact load, the related damage is likely to occur. In the following section the main criteria for damage onset in foam core sandwich structures are briefly presented, more theoretical details can be found in [55, 56, 61, 62].

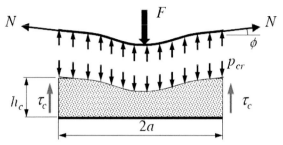

Figure 22. Balance of forces in a region with crushed foam core [56]

Core crushing initiates already at very small deflections of the face sheet, so that small deflection plate theory can be applied. It can be assumed that the foam behaves elastically prior to crushing onset, which may be estimated by considering the small elastic skin deflection as a plate on elastic foundation. The elastic foundation is given by the stiffness and thickness of the core and the face sheet. Figure 22 illustrates the vertical equilibrium of a foam core sandwich section with a core thickness h_c, a skin thickness h_f and a crushing zone with a radius a being indented by a hemispherical impactor with a radius R_{imp}. The critical loads for core crushing can be estimated as follow [56]:

$$F_{cr} = 4p_{cr}\sqrt{Q_f^* h_f^3 h_c/(4.14 Q_{cz}^*)} \qquad \text{for} \qquad h_c \leq h_{cmax};$$

2.22

$$F_{cr} = 3\sqrt{3} p_{cr}[Q_f^* h_f^3/(6Q_{cz}^*)]^{2/3} \qquad \text{for} \qquad h_c > h_{cmax}.$$

With p_{cr} is the crushing stress of the core, and Q_f^* and Q_{cz}^* are the effectives stiffnesses of the skin and the core respectively. The expressions for the calculation of the effective stiffness are available in the appendix A1.
h_{cmax} can be calculated as follow:

$$h_{cmax} = h_f \frac{32}{27} (\tfrac{4}{3} Q_f^*/Q_{cz}^*)^{1/3}.$$

2.23

The critical load for skin delamination onset F_{d1} may be written [55]:

$$F_{d1} = \pi \sqrt{8Q_f^* h_f^3 G_{IIc}/9}.$$

2.24

Where G_{IIc} is the fracture toughness of the face sheet in mode II.
The delamination occurs at very low loads and leads to the loss of the skin bending stiffness while maintaining the membrane action. The delaminated skin may be modelled as a membrane under equilibrium wrapping around the hemispherical impactor. By increasing the load the membrane fails due to tensile skin rupture when the ultimate tensile failure strain ε_{1t} is reached. The skin rupture load F_r is expressed by [56]:

$$F_r = 4\pi R_{imp} h_f E_r \varepsilon_{1t}^2/(1 - \vartheta_r).$$

2.25

Where E_r and ϑ_r represent the average Young´s modulus and Poisson´s ratio of the skin and are equal to the corresponding properties of a quasi-isotropic laminate.

To calculate the radius of the zone under the impactor with core crushing, the equilibrium between the contact load F, the core crushing reaction and the skin membrane load N at the edge of the crash zone is set [56]:

$$p_{cr}\pi a^2 = F - 2\pi a N\theta/F. \qquad 2.26$$

The dimensionless squared radius of the crushing zone can be written following [56]:

$$\bar{a}^2 = \frac{p_{cr}\pi a^2}{F}. \qquad 2.27$$

It describes the fraction of load carried by the skin membrane action during indentation. For practical use it can be assumed that $\bar{a}^2 \leq 0.8$. This assuption is true where membrane model assumptions are fulfilled (core shear cracking takes place at high loads and face sheet deflections).

Under consideration of the classical sandwich panel theory the entire out-of-plane shear load is uniformly carried by the core at the shear stress τ_c. For the case of vertical equilibrium in the crushed core region the following core shear stress for a given impact force F [56] is estimated following this relation:

$$\tau_c = \frac{1}{h_c}\sqrt{\frac{F\bar{a}^2 p_{cr}}{4\pi}}. \qquad 2.28$$

The maximum achievable shear stress is limited by the skin rupture load and can be obtained by the insertion of F_r and the associated crush radius a in the previous equation [56]:

$$\tau_{c,max} = \frac{\varepsilon_{1t}}{h_c}\sqrt{\frac{R_{imp}h_f E_r \bar{a}^2 p_{cr}}{(1-\vartheta_r)}}. \qquad 2.29$$

To obtain the critical impactor load for core shear cracking the shear stress τ_c is replaced by the shear strength of the core $\bar{\tau}_c$ and is given by [7]:

$$F_{cs} = \frac{4\pi\bar{\tau}_c^2 h_c^2}{\bar{a}^2 p_{cr}} \geq \frac{4\pi\bar{\tau}_c^2 h_c^2}{0.8 p_{cr}}. \qquad 2.30$$

It is important for a designer to avoid core shear cracking. This problem can be solved by having a sandwich configuration with a critical load for core shear cracking higher than the load for skin rupture onset. The transition between the two limit cases is given when both critical loads are equal and can be written as [7]:

$$\frac{R_{imp}h_f E_r \varepsilon_{1t}^2}{(1-\vartheta_r)} = \frac{\bar{\tau}_c^2 h_c^2}{\bar{a}^2 p_{cr}}. \qquad 2.31$$

The minimum core thickness for a given face sheet thickness is obtained by reorganising the previous equation to [7]:

$$h_{c,min} = \frac{\varepsilon_{1t}}{\bar{\tau}_c}\sqrt{\frac{R_{imp}h_f E_r \bar{a}^2 p_{cr}}{(1-\vartheta_r)}}. \qquad 2.32$$

2.4 Chapter summary

Sandwich structures have a high specific strength and stiffness compared to other materials and gives the possibility to save weight compared to monolithic composites. While sandwich structures are widely used in load bearing applications especially in rotor blades and high performance boats, its use in civil aerospace applications is limited to secondary aircraft structures and cabin applications. The limited use in civil aircrafts is due to the susceptibility to out-of-plane loading, for instance runway debris and tool-drop, and the strict damage tolerance requirements.

Honeycomb cores are the most used sandwich structures in aerospace applications due to better lightweight performance compared to cellular foams. But, they have drawbacks like water accumulation in the cells and expensive manufacturing. These drawbacks are overcome by closed-cell foam cores like PMI-foams, which offer better shaping of the sandwich structure and more efficient manufacturing. To enhance the mechanical properties of foam core sandwich dry fibre pins are inserted into the foam. It is reported that the impact resistance and the damage tolerance are improved. The pins can be inserted under different pattern and inclination angle, which enables a customised shaping of the mechanical properties of the core.

To understand the behaviour of structure during impact, the impact responses can be classified into two main types: ballistic impact response and non-ballistic impact response. The contact time during impact is used to distinguish between these two types. It was found that the mass criterion should be used to characterise the non-ballistic impact response, therefore we can differ between small mass impact, large mass impact and an intermediate case. The small mass impact response is local and dominated by flexural waves, while the large mass response is global and has a quasi-static nature.

Core shear cracking is the most critical failure mode of foam core sandwich structures. For impact resistant structures this damage mode should be avoided. For the most known damage types, core crushing, delamination face sheet rupture and core shear cracks; criteria for damage onset are available. Since the critical load for core shear cracking is limited by the load for skin rupture, it is advised to design foam core sandwich structures with critical load for core shear cracking higher than the skin rupture load.

3 Experimental characterisation of the Tied Foam Core-sandwich structure

In this chapter the results of quasi-static tests performed on pin-reinforced sandwich specimens are presented. The investigation of the quasi-static properties of the Tied Foam Core (TFC) sandwich is fundamental to generate material input data for the simulation models and to understand the damage behaviour under different loading conditions. Flatwise compression tests, out-of-plane shear tests are fundamental tests to characterise sandwich structures. They were performed and analysed in addition to foam and sandwich indentation tests. Furthermore, the damage modes for every test were investigated using the X-ray computed tomography. The mechanical behaviour as function of the reinforcement parameters was explored and evaluated. The available literature related to each performed test was reviewed in the corresponding sections.

3.1 Materials and manufacturing

All tested specimens in this chapter consist of two thin carbon-epoxy face sheets separated by a low-density Polymethacrylimide (PMI) foam core. Rohacell foams are a mixture of a copolymer of methacrylonitrile (C_5H_5N), methacrylic acid ($C_4H_6O_2$) and few key additives including alcohol as foaming agent. They are produced by foaming and thermal expansion of solid copolymer sheets in large ovens at high temperatures [63]. They can reach a heat resistance up to 210 °C. A vacuum assisted resin infusion process for sandwich structures based on the process described in the previous chapter was used to manufacture all the sandwich panels in this work. In order to avoid the creation of air bubbles in the pins, two resin inlets were used; one for face sheet in the bottom and one for the top face sheet, and the impregnation of the upper face sheet was started with a small delay to the impregnation of the face sheet in the bottom of the sandwich panel. This enables to evacuate the air properly from the dry fibre pins before the connection of both resin fronts, as the resin flows from the bottom to the top of the pins and meets the resin front of the top face sheet. The modification of the manufacturing process enables to have a reproducible quality of the pins. The used vacuum infusion build up is depicted in Figure 23.

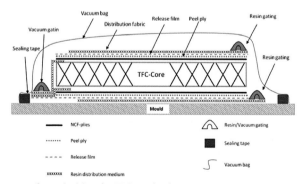

Figure 23. Vacuum infusion build up for TFC-sandwich

The face skins were made of Toho Tenax HTS40 carbon fibre Non-Crimp Fabrics (NCF) impregnated by Hexcel RTM6 epoxy resin. The mechanical properties of the face sheets materials are depicted in the manufacturers' data sheets [64, 65]. The skin faces were made up of four triaxial NCF-layers, with the layup (45°/0°/-45°) and (-45°/0°/45°), symmetrically stacked, which resulted in a face sheet thickness of about 1.5 mm. The closed cell PMI foam ROHACELL® 71 HERO [24] with a density of about 75 kg/m^3 and a thickness of about 25.7 mm was used as core material. Figure 24 shows the used sandwich lay-up. For the reinforcement of the foam core, two types of fibres were used: T800H carbon fibre from TORAYCA [66] and 1383 Yarn glass fibre from PPG Fiber Glass [67]. The cured pins have a diameter of about one mm. All the specimens tested in this work have been cut using a diamond cutting disc saw with a reduced feed velocity to avoid excessive pins and foam damages at the cutting surfaces.

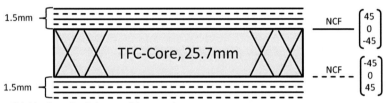

Figure 24. Used sandwich lay-up of the tested specimens

3.2 Tested TFC-configurations

In this work, four different pin-configurations have been investigated. All configurations have the same pin-pattern and a stitching angle of 50°. In addition, sandwich specimens without reinforcement were tested as reference configuration. Figure 25 shows the used pin-pattern and a 3D-view of the unit cell. Every unit cell has four pins inserted with an inclination angle ϕ of 50° to the top panel surface. The unit cells are placed to form a predefined pattern. The pin-density is controlled through the unit cell distance (distance between two aligned pins in a unit cell) and the distance between two unit cells (distance between two aligned pins with the same orientation).

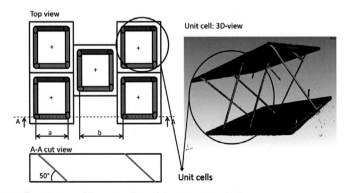

Figure 25. Different views of the unit cells and the pin pattern [68]

Table 1 gives the nomenclature and the unit cell parameters of the investigated TFC-configurations. Glass and carbon fibre pins and two different pin volume contents have been used. The pin-volume fraction (PVF) of the configurations CFRP.10 and GFRP.10 is about four times higher than that of the configurations CFRP.20 and GFRP.20. The PVF was calculated based on CAD-Data and TFC-core geometry of about 50 x 50 x 25.7 mm, only full embedded pins were considered.

Table 1. Unit cell parameters of the TFC-configurations

Configuration	Pin material	Unit cell width	Distance between two unit cells	Pin volume fraction [%]
CFRP.10	CFRP	10	20	0.74
CFRP.20	CFRP	20	40	0.20
GFRP.10	GFRP	10	20	0.74
GFRP.20	GFRP	20	40	0.20
H.71	Reference configuration without pins			0

3.3   Damage analysis with X-ray computed tomography

a)   b)   c)

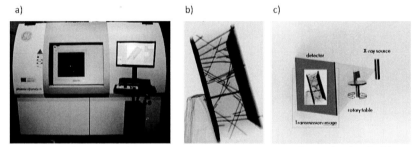

Figure 26. (a) The tomography computer Phoenix-X-ray Vtome X m (research edition); (b) X-ray image of the inspected part; (c) CT operation principle with a specimen having a tilt against the Rontgen cone beam

Instead of using a chemical process or high speed water jet stream for foam removal or analyzing micro-sections by light microscope, a computer tomography is used for a non-destructive inspection of pin reinforced sandwich structures. This can be done either by using virtual cross sections out of the 3D data or by a radiographic image itself (Figure 26-b). To receive the 3D data the test object is positioned on a precision rotational stage that is between the X-ray source and X-ray detector, images are acquired during the rotation at constant steps. The scan usually covers a rotation of 360 degrees and depending on the step size, a varying number of images are thus collected (Figure 26-c). The imaging system produces a two-dimensional shadowgraph image of the specimen. From a series of 2D shadowgraphs, the reconstruction software provides 3D volume results using filtered back projection algorithm. The volume CT data are rendered as voxels with three-dimensional resolution below 1 micrometer (Nano-CT) to hundreds of micrometers depending on X-ray detector pixel, X-ray spot, and object size. In this work the software Volume Graphics has been used to analyse the 3D-data.

For the inspection the micro focus X-ray tube (240 kV / 320 W) at a voltage of 110 kV and 210 μA was used. Within a so-called "Fast-Scan" 2024 x 2D images on a 360° rotation where taken with about eight minutes measuring time. The voxel size equals 23 μm. In Figure 26-b the 2D view of a tested specimen is displaced. Since there is a high difference regarding the density of pin, foam and the Carbon-fibre-Reinforced plates, the foam can virtually be separated, giving a free view on the pins. To get a detailed overview on the pin failure virtual cross sections are taken (Figure 34), whereby the separation of foam and pins can be seen.

3.4 Flatwise compression test

In this section the results of experimental investigation of TFC-sandwich specimens tested under compression loads are presented. The effect of pin material and pin volume fraction on the compressive properties and the energy absorbency of the sandwich structure have been investigated. The failure modes of the pins were determined using the X-ray computed tomography described in the previous section. Moreover, an analytical model to predict the compressive properties of the pin reinforced sandwich has been tested and evaluated.

3.4.1 Literature review

Many researchers [69-71] have intensively investigated the influence of the pin reinforcement on the compressive properties and the energy absorption capacity of the foam core sandwich structure. The most common conclusion of these investigations is that, independent from the pinning process and concept the compressive properties and the crushing energy of the foam core sandwich are enhanced when composite pins are used. Marasco investigated in his PhD thesis [72] the effects of the presence of foam material, pin insertion angle and pin areal density on the compressive properties and failure behaviour. The compressive performance of the pin reinforced foam core sandwich was compared to the performance of Nomex® and Al-Honeycomb samples with similar density. He concluded that the Nomex® honeycomb outperforms the tested X- and K-Cor samples regarding the compressive strength and absorbed energy, while the Al Honeycomb showed an enhanced energy absorption capacity. Considering the effect of the foam material Marasco spotted the synergistic effect between the foam and the pins, as the compressive strength of the pin reinforced sandwich is higher than the sum of the contribution of the foam and the pins alone. The foam creates a lateral support to the pins, so that the critical stress for the elastic buckling of the pins is increased and the compressive strength of the sandwich specimen is notably improved. Lascoup et al. [12] investigated the behaviour of angled stitched sandwich under different loading conditions. They focused on the effect of stitch step and angle on the mechanical performances. They highlighted that the mechanical properties are generally improved with the increase of stitching density and the through-the-thickness compression properties are enhanced by stitching vertically. Fleck and Cartié [73] compared the failure behaviour of Titanium and carbon fibre pins produced with the X-Cor pinning technology under compression loads and underlined the same synergistic effect between foam core and pins highlighted by Marasco [72]. Nanayakkara et al. [37] analysed the mechanical behaviour and failure modes of sandwich specimens with vertically inserted Z-pins. They described a significant improvement of the compressive properties with the increase of the pin volume fraction. The failure occurrence was acoustically monitored during the tests and they concluded that pins began to fail in the elastic regime and continued to fail after reaching the yield point. X-ray computed tomography showed that longitudinal splitting, kinking, crushing and fragmentation were the main failure modes of the pins.

3.4.2 Test description

In order to highlight the effects of pin-configurations on the compression performance of sandwich structures in term of compressive modulus E_{cz}, compressive strength $\bar{\sigma}_{cz}$ and absorbed crushing energy E_{cr}, flatwise compression tests based on DIN 53291 Standard [74] were performed. The specimens had a thickness of about 28.7 mm (inclusive skins), a width and a length of about 50 mm both. The compression load was equally applied over the contact surface of the test specimen using a 250 kN Zwick universal testing machine. A constant displacement rate of 1mm/min was maintained until 10% compressive strain, and then increased to 2 mm/min until a compressive strain of about 75% was reached. The reaction force and the crosshead displacement were continuously recorded. Five samples from each configuration were tested. The test setup is illustrated in Figure 27.

Figure 27. Test specimen under compressive load

The compressive elastic modulus E_{cz} was calculated by using a best fit to the linear region of the stress strain curve, so that E_{cz} was obtained from the curve gradient using the following equation [74]

$$E_{cz} = \frac{\Delta\sigma}{\Delta\varepsilon} \, . \qquad\qquad 3.1$$

Where $\Delta\sigma$ and $\Delta\varepsilon$ are the stress and strain differences in the considered linear region respectively. The compressive stress was calculated from the applied load P and the initial cross section surface of each specimen A_c as follow [74]

$$\sigma_{cz} = \frac{P}{A_c} \, . \qquad\qquad 3.2$$

The compressive strain is defined as [74]:

$$\varepsilon_{cz} = \frac{\Delta h}{h_c} \, . \qquad\qquad 3.3$$

Where Δh and h_c are the crosshead displacement and the thickness of the foam core respectively. The compressive strength $\bar{\sigma}_{cz}$ is measured at the yield point, which is the maximum stress after the linear region of the stress strain curve. The absorbed energy E_{cr} during crushing was calculated from the area beneath the load displacement curve. Only the area up to 60% compressive strain was considered to remove the effect of foam densification due to core crushing and compaction, at which the reaction force rash increases.

3.4.3 Results and discussion

3.4.3.1 Effect of pin material and pin volume fraction

The average stress-strain curves of the pin-reinforced configurations and the reference sandwich without pins are presented in Figure 28. All specimens show the typical slope of a compressive stress-strain curve of foam core material. In the elastic region, the load increases linearly until reaching the compressive strength. For specimens without pin reinforcement the load peak means that the foam core cell walls begin to collapse. For TFC-specimens the pins offer an alternative load path and the structure continues to carry loads higher than the collapse load of traditional sandwich structures until the pins begin to fail and the load carrying ability is degraded. The failure of the pins is characterised by a load pick after the linear elastic region. An increase of the pin-density led to a better improvement of the compressive strength and to more pronounced load drop after first pin-failure compared to the unreinforced specimens. Afterwards the collapse plateau region begins and the compressive stress rises slowly with increasing strain. This region is mainly dominated by the crushing of the cell walls, since the carrying ability of the pins is dramatically reduced. However, the stress level in the collapse plateau region for the specimens with pin-reinforcement still higher compared to the reference configuration, then the collapse of the pins continued in this region and they still have reduced load carrying ability. At about 60% strain, the start of the densification region is reached and the stress begins to increase rapidly. From about 30% strain until the end of the experiments the stress-strain responses of the configurations with pin-reinforcement, except configuration GFRP.10, is nearly the same, which approves that after complete collapse of the pins the sandwich response is mainly controlled by the foam core behaviour. Since the breakage behaviour of glass fibre is more ductile than that of carbon fibre, glass fibre pins showed smoother damage behaviour and provided improved mechanical performance after reaching the compressive strength of the structure.

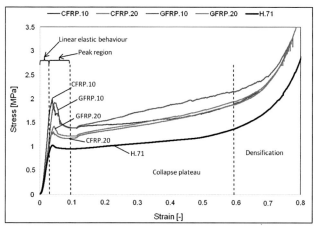

Figure 28. Average Stress-Strain response of the tested configurations

From the observation of the compressive moduli of the tested configurations depicted in Figure 29, is clear that adding pins even in small quantity increases the stiffness of the sandwich material. Although the specimens CFRP.20 and GFRP.20 have different pin-materials, they have nearly the same compressive modulus. That means the effect of pin-material at low pin volume fraction level is not relevant for the compressive stiffness of the structure. When the pin volume fraction is quadrupled, specimens with CFRP-pins show a higher increase of the structure stiffness compared to specimens with GFRP-pins. It seems to be logical since the stiffness of CFRP-pins is superior to that of GFRP-pins and the pin-influence on the mechanical properties of the structure is more significant at this pin volume content level.

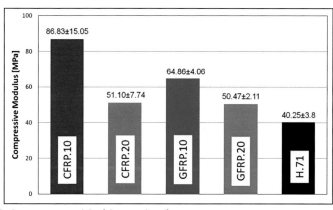

Figure 29. Compressive moduli of the tested configurations

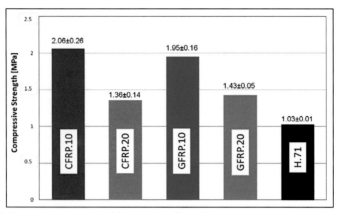

Figure 30. Compressive strengths of the tested configurations

Regarding the compressive strengths of the tested configurations illustrated in Figure 30, increasing the pin volume fraction enhances the compressive strength, which is doubled with the configurations CFRP.10 and GFRP.10. Unlike the behaviour of the compressive modulus, the compressive strength seems to be more independent of the tested pin-material, since specimens with the same pin volume fraction but different pin-materials have almost the same compressive strength.

Another key parameter to compare the different configurations is the absorbed crushing energy during compression. Since energy absorption occurs mainly in the collapse plateau, only the area beneath the load-displacement curve up to 60% compressive strain was considered. The absorbed crushing energies per unit length of the tested configurations are diagrammed in Figure 31. A maximum crushing length of 60% of the specimen high was considered. The introduction of pins into the foam core even with small amount increases generally the absorbed crushing energy compared to unreinforced foam. The configurations CFRP.20 and GFRP.20 led to an increase of about 35% of the crushing energy per unit length compared to the reference configuration. More pins in the foam core led to an enhancement of the energy absorbency and reached 58% improvement for configuration GFRP.10. This increase of absorbed crushing energy for specimens with pin reinforcement can be explained through the additional energy absorption due to pin breaking and crushing. At high pin density level, specimens with glass pins absorbed about 10% more energy compared to specimens with carbon pins, because of the better performance of glass fibre pins in the collapse plateau region. Otherwise, specimens with low pin density show the same energy absorbency, independent of pin material. These observations confirmed that the effect of pin material is only relevant when the pin volume fraction is high enough. This conclusion is only valid for glass fibre and carbon fibre pins tested in this work.

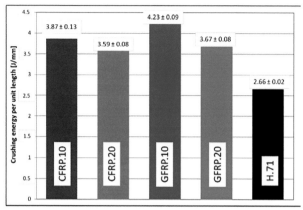

Figure 31. Absorbed crushing energies per unit length of the tested configurations

The last qualitative characteristics to evaluate the different configurations are the standard deviations of the calculated material properties and the fluctuations in the stress-strain response. From the properties highlighted above is to notice that the increase of pin volume fraction or using carbon fibre pins instead of glass fibre pins increased the standard deviation of the determined properties and the fluctuations in the stress strain curves illustrated in Figure 32. That means, increasing the difference of stiffness and strength between the core and pin materials leads to a growth of fluctuations and reduces the repeatability of test results. This can be explained through the increase of the inhomogeneity of the compound material.

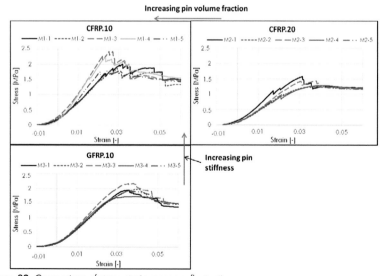

Figure 32. Comparison of stress-strain response fluctuations

3.4.3.2 Effect of specimen geometry on compressive properties

Since the geometry of the test specimen described in DIN 53291 [74] does not consider the pin reinforcement and was elaborated for standard foam and honeycomb sandwich, the effect of specimen cross section on the compression performance of pin reinforced sandwich was investigated in this work. Specimens with about 100 x 100 mm cross-section area and same thickness like the reference configuration were tested. The compressive strength and modulus were then calculated and compared to the previous results. In Figure 33 the average stress-strain curves of the 100 x 100 mm-specimens are compared to the average stress-strain curves of the norm compatible samples. The development of the curves shows very similar behaviours in the linear, collapse plateau and densification regions. The yield stresses of specimens with 100 x 100 mm cross-section are higher than that of the norm geometry. Configuration CFRP.10 shows an exception, the stresses in the collapse plateau and densification regions are higher for the specimens with 100 x 100 mm cross section surface. This behaviour could not be explained, since all specimens were manufactured and tested under the same conditions. The measured compression properties depicted in Table 2 show that, both specimen geometries delivered nearly the same compressive modulus. Concerning the compressive strength an increase was generally observed for the specimens with 100 x 100 mm cross section surface. This growth is higher when the difference of compressive strength between pin material and foam is larger. Configuration CFRP.10 has the highest increase of compressive strength compared to specimens with glass fibre pins or lower pin density.

These results show that the norm compatible specimen geometry is generally suitable for pin reinforced sandwich specimens, since the determined material parameters are conservative. When the pin density is high and the compressive strength difference between pin and foam materials is too large, specimens with larger cross-section surface would deliver results that are more accurate, as the standard deviation is smaller compared to the small specimens. For better experimental results it is recommended to use the large specimens to reduce the scatter of the pin volume faction in a sandwich specimen and to improve the accuracy of the extracted material properties.

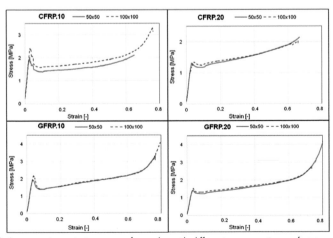

Figure 33. Average stress-strain curves of samples with different cross-section surfaces

Table 2. Effect of specimen cross-section on compression mechanical properties

Configuration	Stiffness [MPa]		Strength [MPa]	
	50x50 mm	100x100 mm	50x50 mm	100x100 mm
CFRP.10	86.83 ± 15.05	88.17 ± 7.42	2.06 ± 0.26	2.51 ± 0.13
CFRP.20	51.10 ± 7.74	50.15 ± 7.60	1.36 ± 0.14	1.36 ± 0.08
GFRP.10	64.86 ± 4.06	61.89 ± 4.11	1.95 ± 0.16	2.21 ± 0.01
GFRP.20	50.47 ± 2.11	48.70 ± 2.08	1.43 ± 0.05	1.52 ± 0.04

3.4.3.3 Failure modes under compressive loads

To evaluate the damage behaviour and determine the failure modes of the tested specimens visual inspections and X-ray computed tomography were performed. Five main failure modes of the compressed specimen were detected. Pin pullout and foam damage at
free side surfaces of the specimen (Figure 34-a) were detected visually. Under compression
loads, the face sheet pushed not fully supported pins. This led to the collapse of the bonding between pin and foam material and afterwards the pins were pulled out. In some cases
the bonding between pin and foam did not fail, which pushed the foam material to follow
the movement of the pin until the foam strength is reached and a crack propagated in the
foam core starting from the interface between pin and foam material (Figure 34-a).

The X-ray computed tomography revealed three different failure modes. Figure 11-b shows
a pin breakage at the pin foot near the connection between face sheet and the pin. In order to increase the fracture toughness of the interface between foam core and face skins,
the pin extensions are folded, so that they lie in the interface. This operation creates a curvature at the pin base, which leads to stress concentration due to notch effect. During the
test, the pins stood high compression and bending loads until the pin strength at the weakest point (pin base) was reached and breakage occurred. Compression failure and pin
fragmentation (Figure 34-c) were also observed. After pin breakage, a gap was created in
the damaged area of the foam. Further loading of the specimen during the experiment led
to gap growth, crack propagation in the foam material (Figure 34-c and d) and core indentation by the broken pin creating more damage in the structure.

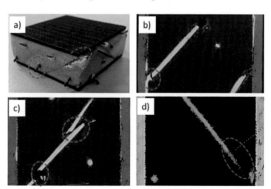

Figure 34. Failure modes of pin reinforced sandwich samples after flatwise compression test: (a)
Pin pull-out and foam damage; (b) pin breakage at pin foot; (c) pin compression failure and fragmentation; (d) widening of the gap between foam core and pin after pin breakage

The load drop after reaching the compressive strength (Figure 28) is a sign of failure occurrence and stiffness degradation of the tested specimen. Since the compressive strength
of pin-reinforced specimens is higher than the compressive strength of the unreinforced
specimens, the steep load drop could lie on the occurrence of pin failure. When the pins
fail, the load carrying ability of the test specimen is dramatically reduced. For the purpose
of identification of the pin failure mode responsible for the load pick when the yield point
is reached, two test specimens were loaded until 5% compressive strain then unloaded. At
this load level, the compressive strength was reached and the load drop has already occurred. Afterwards the test specimens were investigated using X-ray micro computed tomography. The micro tomography revealed that pin breakage at pin base (Figure 35) is the
first pin failure mode that occurred after reaching the compressive strength of the test
specimen. The same breakage location was also observed by Lascoup et al [12] and Fleck
and Cartié [73] in their investigations. To understand the cause of the occurrence of pin
breakage at the above-mentioned location, the loading situation of the pin is analysed as
follow: the compressive load of the testing machine is introduced into the core perpendicular to the skin surface. It is assumed that the pins are fully clamped at the ends, so that it
represents a rigidly supported beam with no rotation allowed. The pin inclination leads to
the splitting of the pin load into two force components (Figure 36): a compressive component acting along the pin and a bending load component acting perpendicular to pin axis.
Since the pins are fully constrained at pin ends, the bending force leads to a stress concentration close to the pin-face sheet connection. The compression forces create an additional
stress in the pin. Due to notch effect created by the flattened pin extensions bonded to the
face sheet, the stress concentration is increased at pin ends, a weak zone is created and
the pin strength is extremely reduced.

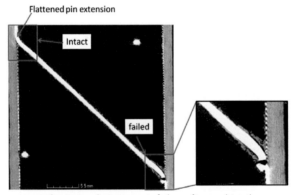

Figure 35. X-ray computed tomography image: pin failure after reaching the compressive strength
of the sandwich specimen

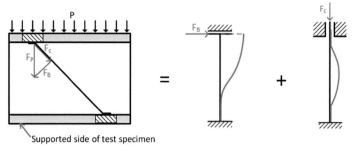

Supported side of test specimen

Figure 36. Forces acting in rigidly supported pins

From the micro tomography it was realised, that all fully embedded pins failed due to pin breakage at pin base when the compressive strength of the test specimen is reached. Moreover, all the pins failed at one extremity while the other extremity was intact, which means that the pin failed first at the weakest location. The elastic buckling failure of the pins was not the dominant failure mechanism like it was observed in the investigations done by Marasco et al [36] and Cartié and Fleck [73].

3.4.4 Analytical method to predict compressive modulus and compressive strength

The analytical model developed by Mouritz [75] was chosen to be tested and evaluated. The model predicts the compressive strength and stiffness of pin reinforced sandwich structures and takes the inclination angle of the pins into consideration. It was assumed that the pin arrangement has no influence on the compressive properties [73]. The compressive modulus of pinned sandwich specimen is calculated using the rule of mixture expression as follow:

$$E_{cz} = E_{fc} \cdot f_{fc} + E_p(\phi) \cdot f_p. \qquad 3.4$$

Where E_{fc} and E_p are the compressive modulus of the foam core and the elastic modulus of the inserted pins, respectively, and f_{fc} and f_p are the volume fractions of foam core and the pins. The compressive strength of the sandwich composite is obtained with the following expression:

$$\bar{\sigma}_{cz} = \sigma_{fc} \cdot f_{fc} + \sigma_p(\phi) \cdot f_p. \qquad 3.5$$

Where σ_{fc} and $\sigma_p(\phi)$ are the compressive strength of the foam core and pin failure stress, respectively. Since no pin buckling failure was observed during the flatwise compression tests and no exact pin strength is available, it is assumed, that the pins fail when the pin compression strength is reached.

The effect of the pin inclination angle ϕ on the compressive modulus and strength of the pins is considered using the Halpin-Tsai equations [75]:

$$E_p(\phi) = \left[\frac{cos^4\phi}{E_{p,x}} + \frac{sin^4\phi}{E_{p,y}} + \left(\frac{1}{G_{p,xy}} - \frac{2 \cdot \vartheta_{xy}}{E_{p,x}} \right) \cdot sin^2\phi \cdot cos^2\phi \right]^{-1}, \qquad 3.6$$

and

$$\sigma_p(\phi) = \left[\frac{cos^2\phi \cdot (cos^2\phi - sin^2\phi)}{\sigma_{p,x}^2} + \frac{sin^4\phi}{\sigma_{p,y}^2} + \frac{cos^2\phi \cdot sin^2\phi}{\tau_{p,xy}^2}\right]^{-1/2}. \qquad 3.7$$

The indexes x and y correspond to the directions along and transverse to the pin axis. $\vartheta_{p,xy}$, $G_{p,xy}$ and $\tau_{p,xy}$ are the Poisson´s ratio, the shear modulus and strength of the pin, respectively. The offset angle ϕ is measured from the compression load direction. It is assumed that the linear elastic behaviour is the same under tensile and compression loading. Since the pins are inserted automatically using the TFC-technology, the orientation angle of the pins is very accurate and the offset deviation is neglected.

The material properties used to verify the analytical model of Mouritz [75] were determined as follow: the mechanical properties of the ROHACELL® 71HERO foam were obtained from the flatwise compression test performed in this study. A compressive modulus of 40.25 MPa and a compressive strength of 1.03 MPa were considered. To improve the accuracy of the analytical model, the manufacturer´s mechanical properties of the glass and carbon fibre were adjusted using laminate software, the rule of mixture and the real fibre volume content of the pins. The fibre volume content of the pins was determined chemically. After manufacturing of the sandwich panels, impregnated and cured pins were retrieved mechanically then grinded using abrasive paper to remove the residual foam cells bonded to the pin surface. Afterwards the chemical extraction was performed. The calculated mechanical properties of the pins are depicted in Table 3.

Table 3. Mechanical properties of the used pins

Property	Glass fibre pins	Carbon fibre pins
Fibre volume content φ [%] (measured)	31.4 ± 0.28	44.2 ± 0.28
Axial compressive modulus $E_{p,x}$ [MPa]	19887	131566
Transverse compressive modulus $E_{p,y}$ [MPa]	4131	4244
Shear modulus $G_{p,xy}$ [MPa]	1534	1822
Poisson´s ratio $\vartheta_{p,xy}$ [-]	0.303	0.284
Axial compressive strength σ_x [MPa]	224	958
Transverse compressive strength $\sigma_{p,y}$ [MPa]	170	196
Shear strength $\tau_{p,xy}$ [MPa]	91	87

The pin volume fraction was determined based on the real test specimen geometry not on the unit cell geometry. That means, the number of fully embedded pins in a test specimen was counted for every pin density using the CAD data and then the pin volume fraction was calculated. It was assumed, that only at both ends fully constrained pins contribute to the increase of the compressive modulus and strength. The pins that were cut during the manufacturing were not considered. This approach resulted in pin volume fraction of about 0.74% for the configurations CFRP.10 and GFRP.10, and about 0.2% for the configurations CFRP.20 and GFRP.20.

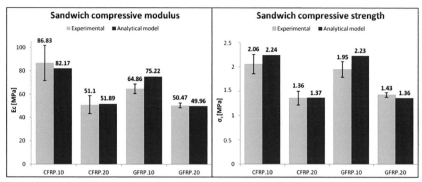

Figure 37. Experimental and analytical compressive properties of the TFC-sandwich

The results of the analytical model in Figure 37 show the same trends observed during the experimental investigations. For low pin volume fraction the model confirmed that the pin material has minor effect on the mechanical properties. Increasing the pin volume fraction led to the improvement of material compressive properties. The analytical model shows that the pin material for high pin volume content has an influence on the compressive modulus, while the compressive strength for the configurations CFRP.10 and GFRP.10 is nearly the same, which was the case in the experimental results. The model gives very accurate results for configurations with low pin volume fraction. At low pin volume fraction the influence of the pin mechanical properties is not significant and the overall mechanical properties are dominated by the properties of the foam core. Increasing the pin volume content increases the failure margin, especially when the stiffness of the pins is approximated and the pin strength is not exactly determined. The analytical model predicts for the configurations CFRP.10 and GFRP.10 higher compressive strengths than the experimental results. This deviation could lie on the assumption that the pins fail when the pin compressive strength is reached, while the X-ray micro tomography showed that the pins failed at pin ends, where a stress concentration was detected. This would make the real pin strength lower than the assumed compressive strength. An improvement in the prediction of the pin strength can be reached through the consideration of a reduction factor in equation 7.

The theoretical results of the model are generally in good agreement with the experimental values of the investigated configurations. The model can be used to design pin reinforced sandwich structures made with the TFC-technology. Increasing the accuracy of the model can be reached through precise determination of the pin stiffness and through a mechanical model for the determination of the pin strength, which considers the complex failure process of the inclined pins. The validity of the model for TFC-Sandwich with other pin inclination angles, other materials or higher pin volume fractions should be further investigated.

3.5 Out-of-plane shear test

In this section the mechanical response of the TFC-sandwich under out-of-plane shear load is investigated. The effect of the presence of the face sheet in the sandwich specimen on the shear properties and the damage evolution is evaluated. Moreover, the influence of the pin material and pin volume fraction on the crack propagation is presented.

3.5.1 Literature review

As previously described in chapter 2.2 shear cracks in the core present the most critical damage mode of a sandwich structure subjected to low-velocity impact. These cracks are invisible and can lead to the loss of the structure integrity. The shear strength of the core material is often the limit stress for core damage initiation. For this reason many researchers have focused in their work on the shear response of pin reinforced foam core sandwich. Independent of the type of the reinforcement an improvement of the shear behaviour has been generally reported and the shear performance improves with the increase of the pin volume fraction in the core [71, 76]. The study performed by Lascoup et al. [12] on foam core sandwich structure with face sheets bonded to the core by the mean of dry fibre stitches shows that the stitch density and stitching angle have a major influence on the shear properties. The best shear performance was reached with a high stitch density and a stitching angle of 45° to the top face sheet plane. However, the studied stitched sandwich showed a brittle damage character compared to unreinforced foam core sandwich characterised by a low shear strain at failure, which may degrade the damage tolerance properties. Marasco et al. [36] investigated the effect of the pinning method and the foam on the shear performance of the pin reinforced foam core sandwich. The pins were arranged in truss pattern, but they were inserted into the core using two different methods. It was found that the shear properties and the failure mechanism depend also on the pinning concept. In addition, test specimens with removed foam material were tested under shear loading. While the foam provides support to the pins during compression and improves the compressive properties, the effect of the foam on the shear properties of the tested unit cell geometry is negligible, as the shear loads are mainly carried out by the pins. In another study published by Du et al. the effect of the alignment of the carbon fibre rods to the shear loading direction on the shear properties was investigated. Three arrangement types were tested: pins aligned along or opposite to the shear loading, and arrayed in X-shaped configuration. It was concluded that the shear properties and the failure behaviour depend on the pin arrangement. The best shear performance was reached with the X-shape arrangement of the pins. In this study a pin configuration based on the X-shape arrangement was adopted, the X-shaped pins were inserted in and perpendicular to the shear loading direction (Figure 25). This pin pattern was chosen to consider the different shear loading directions of a sandwich panel when used in a real application.

3.5.2 Test description

To determine the shear moduli and strengths of the tested configurations, shear tests in accordance to DIN 53294-standard [77] were conducted. From every test configuration, four test specimens were tested. A test specimen has a length of 390 mm, a width of 100 mm and a core thickness of about 25.7 mm. In order to choose a suited specimen design, two types of test specimens have been tested: Type A is a standard sandwich specimen with face sheets bonded to the steel plates and has the same build up like described in section 3.1. The specimens were bonded to the loading plates by means of the adhesive epoxy system Scotch-Weld® 9323 B/A [78]. Type B is a specimen without face sheets, the pins were impregnated with RTM6 epoxy and the curing took place with the same process parameters for the manufacturing of type A specimens. The foam core with the cured pins was then bonded to the loading plates using the epoxy system Hexion L385/386 [79] reinforced with cotton flocks. The damage behaviour was then analysed and a specimen type was chosen for the interpretation of the results. Figure 38 shows the test setup with all parts labelled.

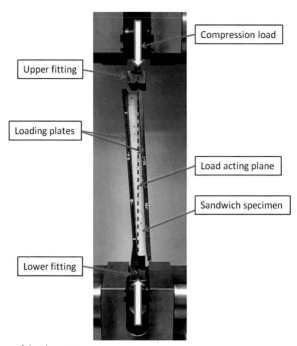

Figure 38. Setup of the shear test

The load was applied to the ends of the rigid plates in compression through a knife-edge bearing at a constant movement rate of the testing machine cross head of 1 mm/min. The load-shear displacement curves were automatically recorded and used to calculate the shear modulus and strength of the core material. Only specimens with core shear damage were considered in the results interpretation, results of specimens with other damage types like adhesive damage or face sheet debonding were discarded. The shear strength was calculated as follow [77]:

$$\bar{\tau}_c = \frac{F_{BS}}{A_0} = \frac{F_{BS}}{l_1 \cdot b}. \qquad 3.8$$

Where F_{BS} is the load at shear failure in N, l_1 the length of the specimen in mm and b is the specimen width in mm.

The shear modulus is determined as follow [77]:

$$G_c = \frac{\Delta\tau}{\Delta\gamma} = \frac{h_c}{l_1 \cdot b} \cdot \frac{\Delta F}{\Delta\vartheta}. \qquad 3.9$$

Where h_c is the thickness of the core in mm and $\frac{\Delta F}{\Delta\vartheta}$ is the slope of initial linear portion of load-shear displacement curve in N/mm. Since the used standard doesn´t contain an explicit definition of $\frac{\Delta F}{\Delta \vartheta}$, it was calculated by using a best fit to the linear region of the load-shear displacement curves.

3.5.3 Results and discussion

3.5.3.1 Selection of the specimen type

The selection of the specimen type was based on two criteria: the damage occurrence and the reproducibility of the results. The stress-strain diagrams in Figure 39 of configuration CFRP.20 show that type A-specimen has less scatter and less non-linearity at the beginning of the test compared to type B-specimen. Moreover, all type A-specimens failed at the same load and strain level.

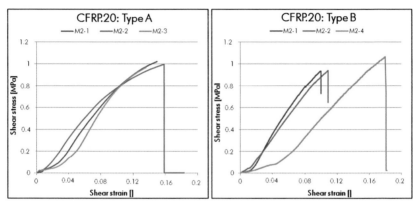

Figure 39. Comparing the stress-strain diagrams of type A and B specimens

Considering the damage behaviour, all type B specimens failed with multiple cracks in the core, which propagate in the adhesive layers. Additionally, the damage pattern was not reproducible and the specimens failed abruptly. However, type A specimens showed the expected damage occurrence, which consists of a shear crack in the middle of the specimen with a defined angle and a propagation in the interface between the face sheets and the core. A comparison of these two damage types is available in Figure 40. This difference in damage behaviour is due to the pin extensions in type B specimens that create a rough adhesive surface and cause an uneven load introduction. However, the load introduction in specimens type A is more homogeneous, as the face sheets remove the unevenness created by the pin extensions and provide a smoother load introduction. In addition, type A specimen has a build-up similar to a real sandwich component, which makes the damage investigation more accurate. For these reasons type A specimen was chosen to investigate the shear properties of the TFC sandwich.

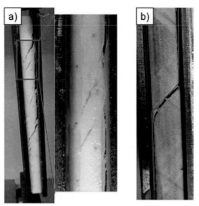

Figure 40. Shear cracks in a) type B and b) type A-specimens

3.5.3.2   Results of specimens test with face sheets

The stress-shear strain curves of the tested configurations are depicted in Figure 41.

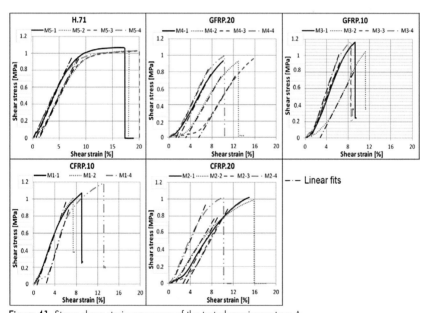

Figure 41. Stress-shear strain responses of the tested specimens type A

Ma-b is the identifier of the test specimen where "a" is the configuration number and "b" is the number of the test specimen in the tested configuration. In every configuration test results of four test samples were considered, except the configuration CFRP.10 where the specimen M1-3 failed due to debonding of the adhesive layer and was not considered in the interpretation.

Comparing the different stress-strain curves shows that using pin-reinforcement increases the non-linearity at the initial portion of the curve compared to the reference configuration without pins. This makes the determination of the shear-modulus of pin-reinforced test specimens more complicated. A toe compensation procedure was applied in order to obtain a zero point on the strain axis (not shown in Figure 41). Moreover, the response of pin-reinforced specimens shows higher results scatter compared to the foam core specimens. Furthermore, the pin reinforcement reduces the plastic deformation of the foam core, which explains the reduced shear strain at failure of the pin reinforced specimens compared to the reference configuration.

The obtained values for shear strength and shear modulus of the tested configurations are summarised in bar graph-form in Figure 42. First observations of the shear properties show that an improvement of shear behaviour is only reached for the configurations CFRP.10 and GFRP.10, which have about four times higher pin volume fraction than the configurations with 20 mm unit cell width. While the configurations CFRP.10 and GFRP.10 led to an increase of shear modulus of about 35% and shear strength of about 15%, the configurations CFRP.20 and GFRP.20 decrease the shear properties. This can be explained by the fact that the pins act at this low pin volume content as defects or notches in the foam material and create stress concentrations, which lead to slight degradation of the shear properties compared to the foam specimens. When the pin volume content was increased to 0.74% (CFRP.10 and GFRP.10), the shear loads were partially carried by the pins, the notch effect was moderated and the shear performance of the sandwich structure was improved. It can be concluded that minimum pin volume content is necessary to improve the shear properties of the foam core sandwich. Considering the pin material, CFRP and GFRP-pins led nearly to the same results when maintaining the pin-pattern unchanged. The latter finding leads to the conclusion that an improvement of the mechanical properties could also be reached when using low-cost pins made of glass fibre.

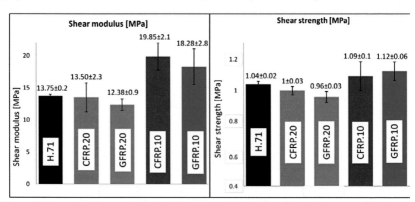

Figure 42. Average values and standard deviations of the calculated shear properties of the tested configurations

3.5.3.3   Failure modes under shear load

The damage behaviour and crack propagation were inspected visually. When the shear strength of a specimen is reached, a load drop in the stress-strain curve occurs and a shear crack with an angle between about 35° and 50° arises and propagates in upper and lower interfaces between the core material and the skins thus the test specimen is divided in two parts. Figure 43 shows an example of a shear crack.

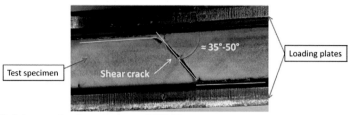

Figure 43. Side view of a shear crack in a GFRP.20-specimen

Regarding the crack redirection and maintaining the integrity of the structure, observations from test execution and failed specimens investigations showed that only specimens with 10 mm unit cell width (CRFP.10 and GFRP.10) prevented the abrupt failure and the collapse of the tested specimen. The other specimens with less pins (GFRP.20 and CFRP.20) and only foam core (reference configuration) showed no residual integrity and collapsed abruptly. A Minimum pin volume fraction (about 0.74%) is mandatory to obtain the expected effects of crack redirection and alternative load path. After crack initiation, the crack propagation is slowed down and the load is mainly carried by the pins. Figure 44 shows the crack redirection effect of two different pin volume fractions. Test specimens with unit cell width of 20 mm (Figure 44-a) show only a minor effect of the pins on the crack propagation and the residual load carrying ability was negligible, while the specimens with 10 mm unit cell width (Figure 44-b) showed a change of the direction of the crack propagation and the pins at the fracture surface assured the integrity of the test specimens through pin bridging effect.

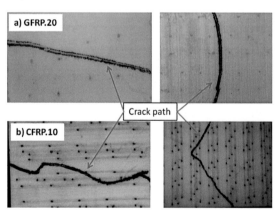

Figure 44. Top views of tested specimens after removing the face sheets: a) GFRP.20-specimen, b) CFRP.10-specimen

Figure 45 shows the crack propagation in thickness direction of GFRP.20 (Figure 45-a) and CFRP.10 (Figure 45-b) specimens. The cracks in the figure are highlighted with red lines. The crack initiated in the CFRP.20 and GFRP.20 specimens (0.2% PVF) with an angle between 35° and 50° to the top surface and continued to propagate in the interface

53

between the face sheet and the foam material splitting the test specimen in two pieces. The two split sections could be detached without remarkable resistance. Increasing the pin volume fraction to 0.74% (CFRP.10 and GFRP.10) led to stopping the crack propagation in the interface between the face sheet and the foam core, as the required propagation energy was already dissipated through crack redirection and pin bridging. As a result the integrity of the specimens was maintained and no splitting was possible. Moreover, no precise range for the crack angle could be defined as the cracks had different orientations at the same time (Figure 45-b).

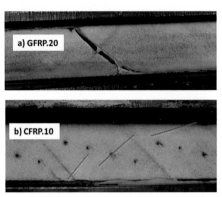

Figure 45. Crack propagation in thickness direction: a) GFRP.20-specimen, b) CFRP.10-specimen

From the results presented in this section an enhancement of the shear properties and improvement of the crack redirection are only possible when a sufficient pin volume fraction is existent. It was expected that the low pin volume fraction of the configurations CFRP.20 and GFRP.20 would lead to a minor improvement of the shear properties compared to the reference configuration without pin reinforcement, but the test results showed that a pin volume fraction of about 0.2% would lead to a slight degradation of the shear properties of the foam core. It is possible that the low pin volume fraction of the configurations CFRP.20 and GFRP.20 was not sufficient to create an additional load path to carry the shear loads and instead the inserted pins created local stress concentrations in the foam material that led to an early collapse of the test samples. Considering the pin-material, GFRP-pins showed, as expected, the same improvement of the shear properties like the CFRP-pins.

3.6 TFC-sandwich indentation test

In this section, the mechanical behaviour of TFC-Sandwich under quasi-static indentation loads has been studied with the aim to investigate the effects of the pin material and pin volume fraction on the indentation response. The pin failure modes were identified using X-ray computed tomography. Studying the indentation behaviour of the TFC-sandwich is important to understand the failure occurrence during low-velocity impact, since the impact load consists of a local indentation load superposed to a global bending load.

3.6.1 Literature review

One of the reasons to use pin reinforced foam core sandwich structures is the improvement of the resistance to local loading like impact and indentation. However, only few studies have investigated the effect of the pins on the indentation behaviour of the sandwich structure. Nanayakkara [80] studied in her PhD thesis the indentation behaviour of foam core sandwich with orthogonal inserted pultruded carbon fibre pins and loaded with cylindrical and hemispherical indenter. The pin loading under the indenter was analysed using the Hertzian contact mechanics. The indentation stiffness and strength were generally improved compared to the unreinforced sandwich structure. She concluded that the efficacy of Z-pins in the stiffening and strengthening of the foam core sandwich is dependent on the loading condition. In another study done by Long and Guiqiong [81] the effect of inclination angle and pinning density on the indentation response of the sandwich structure was studied. It was stated that the effect of inclination angle is marginal compared to pin density and pin configuration. Moreover, it was reported that pin buckling is the main failure mode of the pins, begins at the end of the elastic region and it depends on the location of the indenter. This dependency on the indenter location was also observed in the force-displacement diagrams. In addition, an analytical model to predict the inelastic response after initiation of core collapse was developed based on the principle of minimum potential energy. The compressive response of the core was assumed to be ideal elastic-rigid-plastic without strain rate effects and the buckling stress of the pins was considered to calculate the collapse resistance of the Z-pinning foam core.

3.6.2 Test description

The tests were performed based on DIN 53291 [74] standard. The specimen dimensions were about 100x100x29 mm. The lower face of the specimen was supported by a rigid steel plate to prevent the specimen bending. The indentation load was applied using a hemispherical steel indenter with a diameter of 25.4 mm. A constant displacement rate of 1mm/min was maintained until 12.5 mm indenter displacement was reached. The tests were conducted in a 250 kN Zwick Roell universal testing machine. The reaction force and the crosshead displacement of the indenter were continuously recorded. Five samples from each configuration (Table 1) were tested. The test setup is depicted in Figure 46.

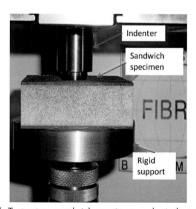

Figure 46. Test setup: sandwich specimen under indentation load

In order to investigate the failure behaviour and to determine the failure occurrence at different load levels, test specimens were loaded to specific load levels and then unloaded. Afterwards the X-ray computed tomography was used to inspect non-destructively the damage in the test specimen. This approach enables to understand the failure occurrence in the sandwich specimen during the indentation test.

3.6.3 Results and discussion

3.6.3.1 Effect of pin materials and pin-volume-fraction

The average indentation load-displacement responses of each configuration are depicted in Figure 47. The overall shape of the indentation curves of the pin-reinforced specimens is similar to the shape of the indentation response of the non-reinforced reference configuration (H.71). Four characteristic regions can be identified in the indentation load-displacement curves:

1. Initial linear elastic region: At this stage, the structure behaviour is linear elastic. After loading, the deformation is totally recovered. This elastic region continues until reaching a load between 800 N and 1100 N, at which the stiffness begins to decrease. This stiffness drop is a sign of damage initiation in the structure. Shortly before this load the linear elastic behaviour ends and the foam crushing starts. The damage at stiffness degradation was investigated with X-ray computed tomography and will be discussed in the next section.

2. Invisible damage propagation: the load continues to increase quasi linearly until face sheet rupture begins. In this region, invisible damages propagate in the indenter region. It is assumed that delamination, interface debonding and pin-breakage are the main failure modes in this region. Small fluctuations in load-indentation curves of pin-reinforced specimens can be identified. Acoustic emissions were also registered.

3. Visible damage propagation: In this region the fibre of the face sheet layers begin to break. This region is characterised by a sharp load drop and loud acoustic emission. After complete penetration of the face sheet the load increases slightly.

4. Plateau region, damage propagation in core material and interlaminar face sheet zone: In this region, the indenter has already penetrated the face sheet and the indentation load is only carried by the core material. The indenter continues to penetrate at nearly constant load. At this stage core crushing in the indentation area, delamination propagation in the face sheet and pin collapse are the main failure mechanism.

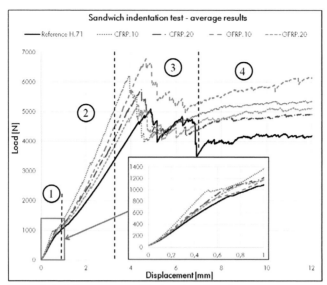

Figure 47. Average indentation responses of the tested configurations

The stiffnesses of the tested configurations in the initial elastic region (region nr. 1) are not the same; the specimens with pin-reinforcement are stiffer than the reference configuration (H.71). The damage initiation occurred almost at the same load level (between 800 N and 1100 N). A significant difference in the indentation response is observed in the second region (region nr. 2). The configurations with pin-reinforcement have higher stiffness compared to the reference configuration. The stiffness increases with the increase of the pin volume fraction and the stiffness of the pins. The pins carry partially the indentation load, create an alternative load path and increase the stiffness of the core. The pin extensions that lie in the interface between the core and the face sheet connect both sandwich facings to each other and create local thickening of the face sheets that leads to the enhancement of the face sheet stiffness and to the increase of the load at face sheet rupture (peak force in region nr. 3). After face sheet rupture the load is mainly carried by the foam and the pins, which raises the load in the plateau region (region nr.4). Configuration GFRP.10 with glass fibre pins and 0.74% pin volume fraction induced the highest augmentation in region three and four. This lies on the high maximum stain to rupture of glass fibre compared to carbon fibre. This leads to the extension of the load carrying capabilities of the pins.

The use of pin-reinforcement leads to the improvement of the quasi-static indentation response of sandwich materials. The stiffness and the load at face sheet rupture are increased especially when enough glass fibre pins are used. The effect of pin material on the indentation response was first observed at a pin volume fraction of 0.74%; at lower pin volume fraction the effect of pin material is minor. The same conclusion was deducted from the test results presented in the previous sections.

3.6.3.2 Failure modes

With the aim to identify the failure mechanism at characteristic load levels, three test specimens were loaded to different loads, afterwards they were inspected using the X-ray computed tomography (one H.71-specimen and one CFRP.10-specimen were loaded to about 2000 N and a CFRP.10-specimen was loaded to about 4000 N).

Figure 48-a and b show delamination in the top face sheet of sandwich specimens in reference configuration H.71 and CFRP.10 after reaching the 2000 N indentation load. The computed tomography approves the critical delamination load of about 1600 N calculated with the equation (2.24) developed by Olsson [55] that predicts delamination limit load. The equation (2.22) for the crushing load of the core delivers a crushing load between 450 N and 650 N for the tested configurations. This load range is slightly lower compared to the load drops in the load-displacement diagrams in Figure 47 that lay between 800 N and 1100 N. The resolution of the CT-scan was not high enough to enable the display of the crushed foam core cells.

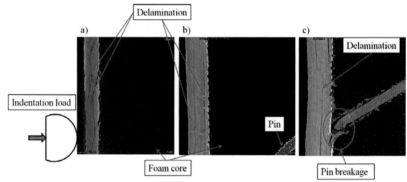

Figure 48. Failure in indented sandwich specimens: a) Reference configuration at 2000N load, b) CFRP.10 at 2000N and c) CFRP.10 at 4000N

Afterwards the delamination continues to propagate in the top face sheet. The pins under the indenter begin to fail at the curved pin extension (Figure 48-c) on the loaded side of the specimen. The same failure was observed during the flatwise compression tests, which confirms that the curved pin extension is a weak point in the pin. This failure mode of the pin differs from the failure observed by Long and Guiqiong [81] who reported that the pins failed due to buckling. This difference may probably lie on the pinning method and the inclination angle of the pins. Long and Guiqiuong [81] used in their investigation sandwich specimens with pin inclination angle between 0° and 20°, which encourages the collapse of the pins under buckling. The pin extensions with folded roving in the TFC sandwich and the pin inclination of 50° lead to pin damage initiation at the curved pin ends. Every pin collapse is reflected in a small fluctuation in the load-indentation curve. The load-indentation response of the reference configuration without pin-reinforcement has no fluctuations between 1000 N and 4000 N, which confirms the observation above. Core crushing under the indenter continues to occur. Further loading of the sandwich specimen after face sheet rupture leads to delamination and face sheet debonding propagation, pin crushing and splitting under the indenter and pin collapse outside the indentation region (Figure 49) due to the bending of the indented face sheet. The nearer to the indentation region, the more the pin crushes and splits. Outside the indentation region, no pin collapse was detected.

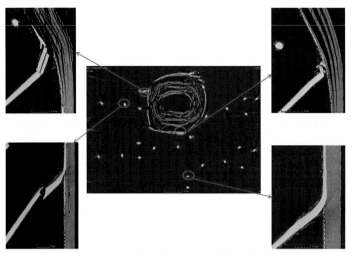

Figure 49. Failure occurrence in pin-reinforced specimen at the end of the indentation test

The performed indentation tests showed that using pin-reinforcement improves the indentation response of foam core sandwich as the stiffness and the face sheet strength are increased. The effect of pin material is first at about 0.74% pin volume fraction evident. The carbon fibre pins led to the highest increase of the stiffness and the glass fibre pins led the highest improvement of the face sheet rupture load. After pin collapse initiation due to bending failure at the curved pin extension nearby the indenter, the pin begins to split and to crush with increased indenter displacement. The identified failure modes give a clear understanding of the damage sequence during low-velocity impact. The similarities between indentation and impact behaviour will be discussed in the next chapter.

3.7   Core indentation test

The aim of this test series is not to generate new material parameters but to provide experimental results for the validation of the simulation model of the pin reinforced foam core. The generated experimental data permit to validate the core material model under complex loading condition and improve the accuracy of the impact simulation.
Since the core indentation test is not a standard test for the characterisation of sandwich materials no actual studies dealing with this topic were found, therefore the literature review is discarded in this section.

3.7.1   Test description

The core indentation test was performed under the same conditions of the sandwich indentation test described in the previous section. The only difference is the specimen geometry. The core indentation specimens were smaller, 50x50 mm instead of 100x100 mm, and the top face sheets were removed using a blade saw, thus the pin reinforced foam core is directly loaded by the indenter. The test setup and a specimen top view are depicted in Figure 50. The load-displacement curves were recorded during the test and the damage behaviour was assessed visually and with micro CT-scans.

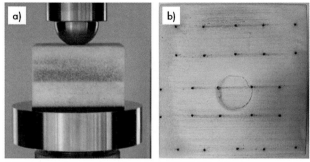

Figure 50. Core indentation test: a) test setup, b) top face of a tested specimen

3.7.2   Results and discussion

The average load-displacement diagrams of the tested configurations are depicted in Figure 51. It can be seen that the use of pins increases the resistance to the indentation load as the reaction force of the pinned specimens is increased compared to the reference configuration without pins. The pins create local reinforcements of the core, which increase the indentation resistance. The resistance increase depends on the number and position of the pins in contact with the indenter, which explains the multiple scatters observed in the indentation diagrams of the pinned specimens. These scatters occur when a pin or the foam fails, or when the indenter gets in contact with a pin, which means a sudden change of the indentation stiffness. Considering the effects of the pin material and pin density on the foam indentation response, no remarkable difference can be observed in Figure 51. Only configuration GFRP.10 shows a slightly higher indentation resistance compared to the other configurations. No satisfying justification for this exception could be found. Unlike the flatwise compression test, only a small surface of the core is loaded by the indenter during the core indentation test, so that only few pins are loaded and the effect of the pin material and density on the core indentation response is limited.

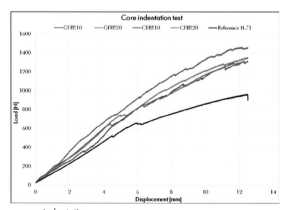

Figure 51. Average core indentation responses

Figure 52 shows the different damage modes observed during the core indentation test. All tested specimens had a circular crack (damage no. 4) with nearly the same indenter diameter. This crack is initiated at high indenter displacement when the failure strain of the foam is reached. When the pin volume fraction is increased, the compliance of the foam is locally reduced, which creates additional cracks under the indenter (damage nr. 5). The micro CT-scans show three different failure modes: pin breakage (damage nr. 1) due to bending deformation of the pin, micro cracks in the core (damage nr. 2) probably initiated by notch effect and residual stresses between the foam material and the pin, and pin-foam interface failure (damage nr. 3).

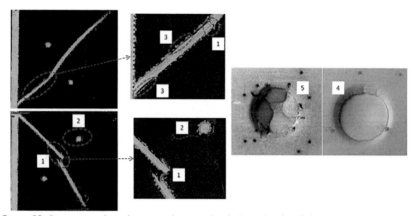

Figure 52. Damage mode under core indentation load: 1) pin bending failure; 2) micro crack; 3) pin-foam interface failure; 4) circular foam cracks; 5) additional cracks

3.8 Chapter summary

This chapter focused on the characterisation of foam core sandwich structure pin-reinforced with the tied foam core technology. Typical quasi-static tests for sandwich structures like flatwise compression, indentation and out-of-plane shear tests were performed. Different influencing parameters, namely the pin material, the pin volume fraction and the specimen geometry have been studied and evaluated. In addition, the damage behaviour under the tested loading conditions was investigated using X-ray computed tomography. This quasi-static analysis of the TFC-sandwich is essential to evaluate the mechanical performance, to identify the damage modes and to generate material input data for the simulation model.

Adding pin reinforcement even at low pin volume fraction (PVF), 0.74% and less, and an inclination angle of 50 ° led to a remarkable improvement of the compressive strength and modulus of the sandwich structure without critical weight penalties. These properties were doubled for the configuration with carbon pins and 0.74% PVF compared to the reference configuration without pins. Increasing the pin density led to a higher enhancement of the compressive properties than using pins with higher stiffness and strength. A difference in compressive properties between specimens with glass and carbon pins was only observed for specimens with 0.74% PVF; specimens with carbon pins had a higher compressive modulus while the compressive strength was nearly the same. Specimens with 0.2% PVF and different pin material had the same compressive modulus and strength.

The pins began to fail at the yield point, where the specimen compressive strength was reached, and the pin failure continued in the collapse plateau region. Pin breakage at pin ends, where the pins are curved and the stress is highly concentrated, was the failure mode with significant consequence on the compressive strength of the pin-reinforced sandwich.

The tested analytical model delivered accurate results especially for the configurations with 0.2% PVF. A higher pin volume fraction led to the increase of the difference between the theoretical and experimental results. An improvement of the accuracy of the model could be reached with more accurate pin properties and especially through the determination of pin strength, which considers the complex stress state at pin base. However, the analytical model is a good tool to choose a suitable configuration of pin-reinforced structure for a specific configuration.

The shear test results presented in this chapter show that an enhancement of the shear properties and crack redirection are only possible when a sufficient pin volume fraction is existent. The configurations CFRP.10 and GFRP.10 led to an increase of the shear modulus of about 40% and 6% increase of the shear strength. This enhancement of the shear properties is low compared to the results reported in the literature [12, 72], it would lie on the low pin volume fraction used in this work compared to other studies. The minor increase of the shear strength lies on the low pin volume fraction of about 0.74%. At this level the pins are not the main shear load carrier, which leads to an insignificant improvement of the shear strength.

It was expected that the low pin volume fraction of the configurations CFRP.20 and GFRP.20 would lead to a minor improvement of the shear properties compared to the reference configuration without pin reinforcement, but the test results showed that a PVF of about 0.2% leads to a slight degradation of the shear properties of the foam core. It is possible that the low pin volume fraction of the configurations CFRP.20 and GFRP.20 was not sufficient to create an additional load path to carry the shear loads, nevertheless, the inserted pins increased the notch effect in the foam and created local stress concentrations in the supporting foam that led to an early collapse of test samples. Similar tendency was observed with crack deviation and integrity enhancement, as only the configurations CFRP.10 and GFRP.10 were capable to enhance the crack deviation and the sandwich integrity. Considering the pin-material, GFRP-pins showed the same improvement of the shear properties like the CFRP-pins, which would reduce the material cost.

Regarding the sandwich indentation tests the results revealed that the overall shape of the load-indentation curve of pin-reinforced specimens is similar to that of unreinforced sandwich. The load-indentation curves of the TFC-sandwich show amplified scatter due to pin breakage and local stiffening of the sandwich. The performed indentation tests showed that using pin-reinforcement improves the indentation response of foam core sandwich as the stiffness and the face sheet strength are increased. The effect of pin material is first at about 0.74% pin volume fraction evident. This finding correlates well with the behaviour observed in the shear and flatwise compression tests.

The carbon fibre pins led to the highest increase of the stiffness and the glass fibre pins led to the highest improvement of the face sheet rupture load. After pin collapse initiation due to bending failure at the curved pin extension nearby the indentation point, the pin begins to split and to crush with increased indenter displacement.

4 Experimental analysis of the impact response of TFC-sandwich at low temperature

In the previous chapter the TFC-sandwich was characterised under quasi-static loading conditions. The effects of pin material and pin volume fraction on the mechanical properties and the damage development in the TFC- sandwich structure were investigated. It was found that the configurations GFRP.10 and CFRP.10 with 0.74% pin volume fraction provide the best improvement of mechanical properties and damage behaviour. For aerospace structures it is important to have good understanding of the damage tolerance at very low temperature, as the material properties of the sandwich components are generally temperature dependent and the thermal loading added to the impact loading would create additional damage to the typical impact damage modes created at room temperature and discussed in section 2.3. In the following chapter the low-velocity impact behaviour of the TFC-sandwich will be investigated with special focus on the temperature dependent impact behaviour. Low-velocity impact tests were performed at room temperature and at -55 °C, and the damage occurrence was identified and analysed. The objective of this chapter is to highlight the effect of the operation temperature on the low-velocity impact behaviour of pin reinforced sandwich structures. A comprehensive and critical review of the published scientific literature dealing with the topic of low-velocity impact behaviour of pinned foam core sandwich structures is presented. A comparison of the indentation results with the impact results at room temperature is also available in this chapter.

4.1 Literature review

One of the most serious barriers to use foam core sandwich constructions in a primary aircraft structure is the insufficient damage tolerance especially after low-velocity impact. Using z-pinning in monolithic laminates shows effectiveness to reduce impact induced delamination and improve the post-impact properties, particularly through crack bridging by the pins [82, 83]. This solution pays off for long delamination, as they allow the creation of a long bridging zone. The same concept could be used in foam core sandwich structures, as the pins would increase the impact damage resistance of the sandwich by means of crack bridging in the core. Many researches have been done to investigate the response of sandwich structures to impact loading and to predict the damage modes that occur. Most of these investigations are summarised in the reviews provided by Abrate [48] and by Chai and Zhu [49]. However, only few researchers investigated the effects of pin reinforcement on the impact behaviour of the sandwich structure. Stanley and Adams [4] studied the feasibility and potential benefits provided by through the thickness stitching of sandwich structures. They investigated the effect of stitch density on the low-velocity impact response and CAI-behaviour. They performed impact tests and investigated the damage occurrence. They concluded that the damage surface decreases and the CAI-strength increases with the increase of stitch density. While only core crushing without cracks was the main failure mode in the unstitched specimens, multiple cracks were observed in the stitched specimens. Lascoup et al. [33] investigated the behaviour of angled stitched sandwich under impact loading. They focused on the effect of stitching and stitching density on the damage behaviour of the foam core sandwich structure. An important improvement of the maximum effort and rigidity was reached. Moreover, compared to unstitched specimens the face sheet debonding was suppressed as the stitches go through both skins and improve the connection to the foam core. Baral et al. [43] compared the structure response of foam core sandwich with PMI foam core and pultruded carbon fibre pins under water impact loading to the response of a honeycomb sandwich structure with the same areal weight and the same facings. They found out that the pin-reinforced foam core sandwich delays the damage to higher impact energies compared to honeycomb structure traditionally used for racing yacht hulls. While most of the available studies reported only benefits of the pin-reinforced foam core sandwich structure compared to the unreinforced foam core sandwich, Nanayakkara et al. [44] concluded that there was nearly no improvement of the impact and post-impact compression properties of the tested z-pinned sandwich. They tested foam core sandwich composite panels with orthogonal pins made of carbon fibre under low-velocity impact and quasi-static flatwise compression loading. On the one hand, the through-thickness stiffness, strength and energy absorption were significantly increased. On the other hand, only small improvement of the structural behaviour at low impact energies and nearly no enhancement of the post-impact properties were registered. Generally, the impact behaviour of foam core sandwich structure is improved when using pin-reinforcement [84, 85], but the effectiveness is strong dependent on the material combination (foam/pin), the pin-configuration (pin-angle, -pattern and density) and how the pins are inserted into the foam core (pinning-technology).

To date, no work could be found in the open literature about the effects of very low temperature on the impact behaviour of pin-reinforced sandwich structures. As the impact behaviour at very low temperature would extremely differ from the behaviour at room temperature, it is important to keep the damage occurrence under control. Gómez-del Río et al. [86] investigated the low-velocity impact behaviour of carbon fibre-reinforced epoxy matrix laminates in low temperature conditions (-60 °C and -150 °C), and they reported that the embrittlement of the matrix combined with the thermal stresses intensified the damage propagation after impact. Block [7] studied in his work the low-velocity impact behaviour of foam core sandwich structure at room temperature and in low temperature conditions (-55 °C), he concluded that the thermal loading leads to the enhancement of shear damage and the creation of vertical cracks in the core.

John et al. [87, 88] investigated different influencing effects that can in- or decrease the residual stresses in unreinforced and pin-reinforced foam core sandwich structures made by vacuum assisted resin infusion process. Depending on the pin-pattern parameters, for instance pin size, distance and pin angle, and the preparation conditions the residual stresses can be minimized. The pin material as well as the thermal process parameters are playing an important role. Negative thermal loading can lead to strains up to 1.3% in the foam core at a temperature of about -55 °C, which would lead to initial degradation of the core. Residual stresses left in the sandwich structure can be estimated at least of about 15% of the overall residual strength left in the foam core, when pins are implemented [87, 88]. The relaxation behaviour of the PMI foam core material should also be considered, as it can reduce up to 65% of the thermal stresses compared to untreated without relaxing PMI foam core sandwich.

4.2   Low-velocity impact: test set-up

Sandwich panels consisting of two thin carbon-epoxy skins and a closed-cell PMI foam core (ROHACELL® 71 HERO) have been manufactured using the Vacuum Assisted Resin Infusion process and the same materials described in section 3.1. The core has a thickness of about 25.7 mm and the face sheets have a thickness of about 1.5 mm with the layup $(45°/0°/-45°)_{2s}$. Three different pin-configurations have been investigated: CFRP.10,

CFRP.20 and GFRP.10, the unit cell properties are described in Table 1. Several impact tests on pin-reinforced foam core sandwich panels have been conducted in the course of this research work. The goal of the performed tests was to evaluate the impact perform-ance at very low temperature and understand the effect of the pin-material and the manu-facturing process on the failure occurrence. As reference, sandwich panels were impacted at room temperature with impact energies between 35 J and 70 J. Afterwards specimens with carbon fibre pins were impacted at -55° and the results were compared to the results of the reference panels impacted at room temperature to enquire the effect of the tem-perature on the impact behaviour. Then, panels with carbon pins were manufactured in autoclave and impacted at -55 °C to show the influence of the manufacturing process on the impact performance at very low temperature. Finally, sandwich specimens with glass fibre pins were tested at -55 °C and the results were compared to the results of the panels with carbon fibre pins impacted at -55 °C with the aim to understand the influence of the pin material on the impact performance at very low temperature. The impact tests were performed at the Fraunhofer Institute for Microstructure of Materials and Systems in Halle (IMWS) using a stationary drop-weight impact testing machine of the type CEAST 9350 (Figure 53). The testing machine has a climate chamber, which can be cooled down to a temperature of about -55 °C using nitrogen gas from a gas tank. After specimen prepara-tion and cooling, the test specimens were placed in the climate chamber and impacted in the centre of the panel by a hemispherical steel impactor with a diameter of 25.4 mm. The test specimens have a dimension of 450 x 500 mm. Two edges of test panel were clamped using steel bars as depicted in Figure 53. Following the mass criteria of Olsson [57] the impact type of the performed tests can be considered as a large-mass non-ballistic impact.

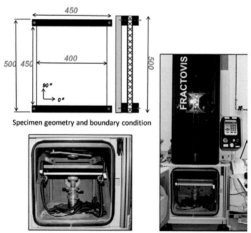

Specimen geometry and boundary condition

Impact test at the Fraunhofer IMWS in Halle

Figure 53. Test setup and specimen geometry [89]

During the impact the contact force and displacement histories, the dissipated energy and the impactor velocity when the impactor hits the panel were recorded. After the impact, the indentation depth was measured and ultrasonic scan was performed. For non-destructive testing of the induced damage (NDT), air-coupled ultrasonic (US) C-scan has been selected. It is capable of detecting face sheet damage within the CFRP and core damage as well as debonding and foam cracks. Micro CT-scans were also performed for selected panels to improve the accuracy of damage detection.

From the force and displacement histories of the impact test combined with ultrasonic C-scan and visual inspection the type of the impact damage could be identified. In case of incertitude the panel should be sectioned or inspected with X-ray computed tomography. Figure 54 shows typical force histories and C-scans of impacted foam core sandwich panels. Face sheet rupture can be identified visually and represents the highest load peak in the force history of the impact. This face sheet rupture load due to impact is mostly identical to the face sheet rupture load determined with the indentation test. In case of delamination without face sheet rupture the damage could be detected with US-scan, as the delamination represents only a small damage surface in the impacted area. In the force history the delamination is assigned to the first significant load drop, which lies in the same load level of the critical delamination load calculated with equation (2.24) in section (2.3.2). The core crushing damage can also be detected with the US-scan and it represents slight stiffness degradation in the linear section of the impact (directly after impact start). The core crushing onset load can be calculated with equation (2.22). Considering the core shear crack the US-scan is one of the most reliable methods to detect it. The shear crack is generally characterised by a large damage surface with an oval shape. In the force history of the impact the shear crack could be usually identified by a sharp load drop lower than the face sheet rupture load. This characteristic load for shear damage can be predicted with equation (2.30) in section. The above outlined characteristic damage loads and C-scan damage patterns are typical for foam core sandwich panels, if pins are inserted in the core the damage loads and the damage behaviour could slightly differ from the damage behaviour of standard foam core sandwich panels.

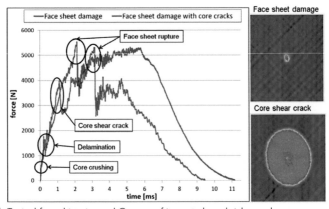

Figure 54. Typical force histories and C-scans of impacted sandwich panels

4.3 Results and discussion

4.3.1 Test results at room temperature

The configurations CFRP.10, CFRP.20 and GFRP.10 were impacted at room temperature with the energies 35 J, 50 J and 70 J respectively. Three sandwich panels for each configuration were tested. The 35 J impact energy is a representative energy for a tool drop scenario. Higher energy are needed to explore the performance of the pin configuration. The ultrasonic inspection was done by an air coupled system with a frequency of 125 kHz. The US-images representing the transmission in percent. The US-scans of the impacted specimens, illustrated in Figure 55, show only local damage in the impacted area for all specimens without exception. Visual inspections and sectioning of the tested sandwich panels show that face sheet rupture, local core crushing and debonding are the only damage modes at room temperature for the tested configurations. No critical shear cracks were observed. The tested sandwich panels fulfil the requirements of damage tolerance and structure integrity at room temperature for the tested range of impact energies, as only local damages were detected.

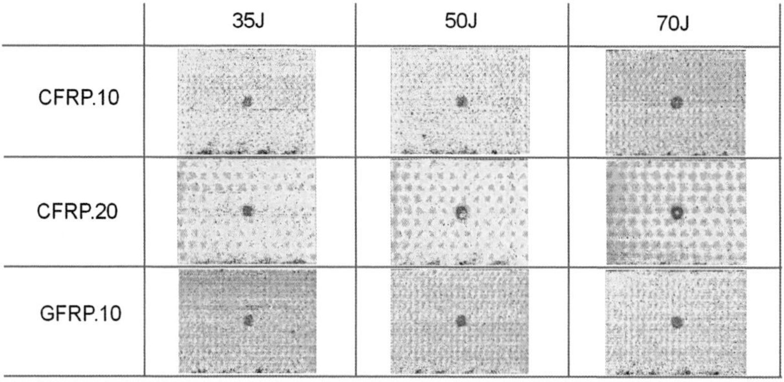

	35J	50J	70J
CFRP.10			
CFRP.20			
GFRP.10			

Figure 55. US-images of the panels impacted at room temperature

In order to inspect the damage behaviour of the pins and the foam in the impacted section a GFRP.10-panel impacted at 70 J was scanned using X-ray computed tomography.

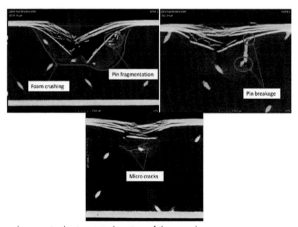

Figure 56. Core damage in the impacted region of the panel

Figure 56 shows limited crushed foam in the region directly struck by the impactor. Moreover, pin breakage and pin fragmentation accompanied with foam micro cracks in the vicinity of the pins can be observed in the same region. Figure 57 depicts the pin damages in the impacted panel section. Pin breakage at pin foot and pin bending breakage are the main damage modes of the pins. Pin compression damage was also observed, but it was very limited compared to the other damage modes. Outside the impacted section of the panel no material failure was observed. The high degree of damage compared to the rest of the panel could be explained by the high deflection and indentation exhibited by this region during the impact event. The 70 J-impact of the GFRP.10-panel was dominated by localised compressive load (no shear cracks in the core), which explains the similarities of damage modes with the damages observed during the indentation test (section 3.6).

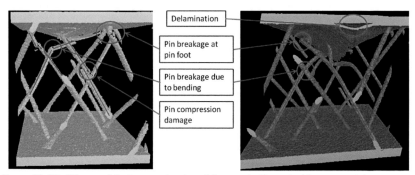

Figure 57. 3D-CT-scan of the impacted region of the panel with masked foam material

In addition to the damage modes, the peak loads at face sheet rupture after impact were compared to the peak loads from the indentation test. For the sake of completeness two panels with the pin configuration GFRP.20 were impacted at 35 J and 50 J and the average of the peak forces was calculated. The panels had the same damage like the panels depicted in Figure 55. The diagram in Figure 58 compares the peak loads from the indentation tests with the average impact peak loads, which end with "-Imp" in the diagram. On the one hand the diagram shows that the configurations CFRP.10 and GFRP.10 with 0.74% pin volume fraction have nearly the same peak load during impact and indentation test (difference of about 120 N), which confirms the strong similarities between the impact behaviour and the indentation behaviour. On the other hand the peak loads from impact are higher than the indentation peak load with a difference of about 1000 N for the configurations CFRP.20 and GFRP.20 with only 0.2% pin volume fraction. This deviation could be explained by the effect of pin volume fraction and pin extension on local stiffness of the face sheet. The pin extensions that lie between the face sheet and the core increases the stiffness of the face sheet, and as result they lead to the increase of the contact force when the pin extension is hit by the impactor. The configurations CFRP.20 and GFRP.20 has a low pin volume fraction, which makes the peak load very dependent on the impact location. As the local stiffened zones cannot be identified during the indentation and impact tests, the deviation between the impact and indentation results cannot be avoided. In case of the configurations CFRP.10 and GFRP.10 the pin volume fraction is high enough to create a homogenous increase of face sheet thickness, so that the pin extensions are always hit by the impactor during impact or indentation tests, which explains the nearly similar peak loads. While the configuration GFRP.10 exhibits the highest increase of peak force during the impact tests, the configurations CFRP.20 and GFRP.20 have surprisingly a peak load slightly higher than the peak load of configuration CFRP.10. This trend differs from the tendency observed during the indentation test. This deviation lies probably on the effect of the pin extension on the contact force of the impactor. It is possible that the impactor hit the local stiffened zones of the CFRP.20 and GFRP.20 panels during the impact test, which led to the strong increase of the peak force.

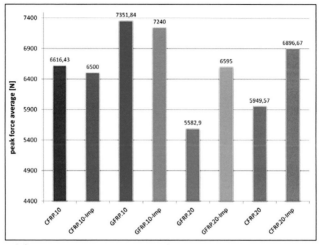

Figure 58. Peak force average: Impact at RT vs. Indentation

4.3.2 Test results at -55 °C using carbon fibre pins

The panels of the configurations CFRP.10 and CFRP.20 were impacted at -55 °C with the impact energies 35 J, 50 J and 60 J respectively. On the one hand, the US-scans in Figure 59 show that the CFRP.10-panels with the highest pin volume fraction have large damage surfaces (magenta surface) with unusual damage shape compared to the typical shape of the damage surface for shear cracks pictured in Figure 54. As the CFRP.10-Panels impacted at 35 J and 50 J exhibited global damage, the impact at 60 J was cancelled. Instead the panel was impact at 35 J. The panel experienced only local damages consisting of face sheet rupture, debonding and core crushing. Hence, the force and deflection histories of these two panels (Figure 60) were compared. The panel with core damage has lower peak force and higher maximum deflection compared to the panel with only local damages. The higher panel deflection is an indicator of additional damages. Considering the force histories, the force development of both panels is identical until the face sheet rupture of the panel with global damage occurred at a load smaller than the force peak of the panel with local damage. No shear crack related load drop before the peak force can be observed in the force history, which strengthen the assumption that these global damages are thermal cracks in the core.

On the other hand, the US-scans of CFRP.20-Panels (Figure 59) have shown only local damage in the impacted area. Reducing the pin volume fraction in the core led to the improvement of the impact performance for the specimens reinforced with carbon fibre, but it has some drawbacks on other mechanical properties like reducing structure stiffness and strength. Compared to the impact results of the reference panels, the thermal loading by the very low temperature of about -55 °C led to the degradation of the impact perform-ance of the CFRP.10-panels, as the foam material becomes more brittle at -55 °C than at room temperature and the thermal loading is superposed to the mechanical loading by the blunt impactor. Moreover, the reproducibility of the impact results was reduced at -55 °C test temperature, as different damage modes were produced for the same impact energy.

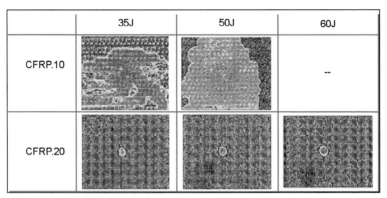

	35J	50J	60J
CFRP.10			--
CFRP.20			

Figure 59. US-images of the panels with CF-Pins impacted at -55 °C

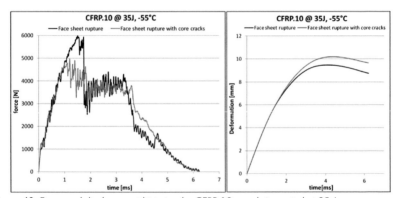

Figure 60. Force and displacement histories the CFRP.10-panels impacted at 35 J

In order to gather more information about the damage occurrence in the CFRP.10-panels, a CFRP.10-panel impacted at 35 J and having global damage mode was investigated using micro-CT-scan. The CT-scan in Figure 61 shows multiple cracks in the foam core. These cracks have an irregular propagation shape without a defined pattern. The crack angle in thickness direction is not constant, vertical and inclined cracks can be distin-guished. Moreover, crack deviation by the pins can also be observed. The pins in the foam core and the dent within the foam core caused by the impact react as a notch that leads to the initiation of foam cracks. Moreover, the difference of the coefficients of thermal expan-sion between the foam and pin material can lead to the creation of process-induced stresses between the pin and the core, as the core and the pin expand differently during the curing at high temperature. This effect is increased when the sandwich is thermally

70

loaded, especially in the case of impact at -55 °C. The notch effect and the residual stresses increase with the increase of the pin volume fraction, which explains the better impact performance of the CFRP.20-specimens. The reduction of the pin volume fraction in the core of the CFRP.20-panels led to the reduction of the process and thermal induced stresses and as a result the improvement of the impact behaviour.

With the aim to improve the impact performance of the pin-reinforced sandwich panels at -55 °C without significant degradation of the structure stiffness, two options are possible. The first option is to prevent the foam expansion during the curing process at high temperature so that the process induced stresses are reduced. The second option is to use other pin materials with higher coefficient of thermal expansion compared to the carbon fibre, thereby the thermal and the process induced stresses are reduced.

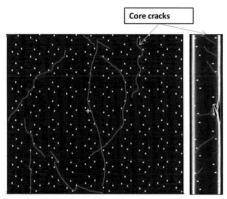

Figure 61. Micro-CT-scan of a CFRP.10-panel impacted at -55 °C

4.3.3   Effects of the manufacturing process on impact damage at -55 °C

To reduce the intensity of the process-induced stresses and evaluate the influence of the manufacturing process on the impact behaviour, CFRP.10-panels were manufactured using the same process parameters used to manufacture the reference panels and described in section 3.1, the resin injection and curing took place in an autoclave oven at a predefined pressure. The autoclave pressure prevents the foam expansion, so that the process induced stress concentrations between the foam core and the pins are dramatically reduced. The manufactured panels were then impacted at -55 °C with the impact energies 35 J, 50 J and 70 J. The maximum impact energy was increased to 70 J with the aim to initiate core cracks.

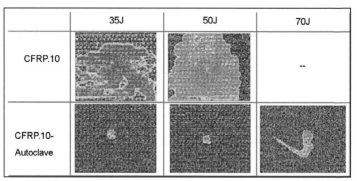

	35J	50J	70J
CFRP.10			--
CFRP.10-Autoclave			

Figure 62. US-images of specimen manufactured in autoclave and impacted at -55 °C compared to CFRP.10-panels with standard manufacturing

Afterwards, the impacted specimens were scanned with air-coupled ultrasound and the results were compared to US-scans of the CFRP.10-specimens manufactured outside the autoclave. The US-scans in Figure 62 show that the specimens manufactured in the autoclave have only local damages, when they were impacted at 35 J and 50 J. No large debonding or cracks in the core were observed. The first crack in the core was observed in the panel impacted at 70 J. Compared to the damage observed in the CFRP.10-panels manufactured with the standard vacuum assisted resin infusion process, the outcome of the damage in the autoclave panels remains limited as only a unique crack in the core and a small debonding surface were observed in the panel impacted with 70J impact energy. Manufacturing the panels in the autoclave oven led to the delay of the critical core damages to higher impact energies compared to the CFRP.10-panels manufactured in standard oven.

These findings lead to the conclusion that process induced stresses have a major influence on the impact behaviour at -55 °C for pin-reinforced foam core sandwich, especially when the difference of coefficient of thermal expansion between the foam and the pin materials is too large. These results approve the conclusions reported by John et al. [90] concerning the influence of the manufacturing process on the residual stresses in PMI-foam core sandwich. A reduction of about 35% of process induced strains was reached, when the curing regime of the conventional VARI was modified.

4.3.4 Effects of the pin material on impact damage at -55 °C

Using pins with a high coefficient of thermal expansion instead of carbon fibre pins is economically more effective to reduce the thermal and process induced stresses than manufacturing the sandwich panels in autoclave oven. In this regard, sandwich panels with glass fibre pins and the same pin volume fraction like the CFRP.10-panels were manufactured with the VARI-process in standard oven and then impacted at the energies 35 J, 50 J and 70 J. The glass fibre with the high coefficient of thermal expansion lead to the reduction of thermal and process induced stresses with a minor decrease of sandwich stiffness and strength compared to the sandwich with carbon pins. The impacted specimens with glass fibre pins (GFRP.10-Panels) were inspected visually and with US-scan, and the damage was compared to the damage of the CFRP.10-panels.

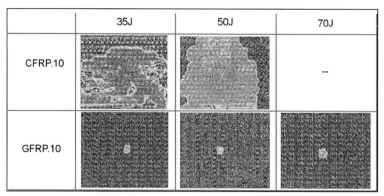

	35J	50J	70J
CFRP.10			--
GFRP.10			

Figure 63. US-images of specimens with GF-pins impacted at -55 °C compared to CFRP.10-panels

The US-images of the GFRP.10-panels shown in Figure 63 display only a local damage in the impacted section of the GFRP.10-panel (the magenta/yellow surface in the centre of the panels). Even at 70 J impact energy no large damage surfaces were observed. Sectioning of the GFRP.10-panel didn´t reveal any cracks in the core. Using glass fibre pins led to the suppression of foam cracks in the investigated range of impact energy ([35 J-70 J]). Using glass fibre pins leads to the reduction of the difference of expansion or shrinkage during the manufacturing process and during the climatisation at -55 °C. As a result, the thermal and process induced stresses between the pins and the foam are decreased, which improves the impact behaviour of the pin-reinforced foam core sandwich panel at very low temperature.

In addition to the embrittlement of the foam material due to very low temperature [91], the residual stresses in the foam core due to the manufacturing process or the thermal loading at very low ambient temperature can intensify the notch effect of the pins, and lead to initial degradation of the foam [92]. Inserting pin reinforcement into the foam core induces residual stresses and strains after curing at 180 °C and cooling down to -55 °C, caused by the prevention of contraction of the foam core in thickness direction. Many factors like temperature, moisture, time as well as cure shrinkage of the resin lead to non-mechanical strains in polymeric materials [88]. The coefficient of thermal expansion (CTE) of the pin material is an important parameter that should be considered during the design of pin-reinforced foam core sandwich. Reducing the difference between the CTE of the pin material and foam material would reduce the process induced and thermal stresses in the core. For this reason panels with glass fibre pins showed better impact behaviour than the panels with carbon fibre pins, as the thermal expansion coefficient of glass fibre is higher than that of carbon fibre. Using aramid fibre pins would not probably improve the impact behaviour compared to the panels with carbon fibre pins, as both fibre types have nearly similar CTEs, that´s why aramid fibre were not considered in this work.

4.4 Chapter summary

In this chapter, the results of impact tests performed on pin-reinforced foam core sandwich panels with different pin-configurations at impact energies between 35 J and 70 J at room temperature and -55 °C were presented. X-ray computed tomography and air-coupled ultrasonic C-scans were used to determine the damages in the core. At room temperature, no critical cracks in the foam core were observed and only face sheet damages were detected. Comparing the results of the room temperature impact tests with the indentation test results shows many similarities regarding pin damage behaviour and critical face sheet rupture load can be identified. The pins failed in both tests mainly due to the loss of bending strength and the face sheet rupture load is the same for the configurations CFRP.10 and GFRP.10.

Impacting pin-reinforced sandwich specimens with carbon fibre pins at -55 °C revealed that the reproducibility of the impact results and the material performance were degraded. Already at 35 J impact energy, the carbon fibre pins led to the initiation of thermally induced cracks in the foam core. These cracks propagate without a defined pattern in the core and create large damage surface with a drastic degradation of structure integrity. In order to improve the damage behaviour at -55 °C new test specimens were manufactured in autoclave to reduce the process induced residual stresses. The autoclave pressure reduces the foam expansion during curing and moderates thereby the process induced stress concentrations. The manufacturing process modification led to the delay of the thermal cracks to higher impact energies. A more efficient solution was reached, when the carbon fibre pins were substituted with glass fibre pins. As a result, the thermal cracks at 70 J impact energy were completely avoided. The glass fibre pins have a high coefficient of thermal expansion.

In addition to changing the impact damage behaviour of the pin-reinforced foam core sandwich structure, the thermal loads lead to the reduction of the peak load at face sheet rupture like depicted in Figure 64. This effect lies on the stiffening of the sandwich materials and on the reduction of the failure strain of the face sheet at very low temperature like reported in [93].

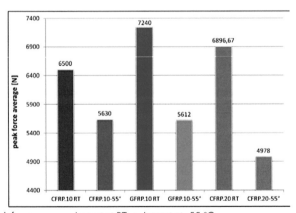

Figure 64. Peak force average: Impact at RT vs. Impact at -55 °C

This chapter shows that thermal loading and process induced stresses have a crucial effect on the damage occurrence in pin-reinforced sandwich structures by impact loading. During impacts at very low temperature, stresses from bending load, thermal stresses and process induced stresses as well as notch stresses are superposed. A large difference of the coefficient of thermal expansion between core material and pin material creates stress concentrations between the pins and the foam material, which may degrade the impact performance at very low temperature. A good choice of the material combination, especially the foam and the pins, enables to avoid the thermally induced cracks and to improve the damage tolerance of the pin-reinforced foam core sandwich structure. Foam materials with a small coefficient of thermal expansion and a high failure strain at very low temperature would improve the damage tolerance dramatically.

5 Numerical simulation of impact on TFC-sandwich structure

In chapter 5 the numerical analysis model to predict the impact behaviour of the TFC-sandwich structure will be developed and validated. In the first section an impact model of a geometric complex sandwich panel is developed and validated. The modelled sandwich panel is a representative section of the side shell of a vertical tail plane of a civil passenger aircraft and the foam core doesn't contain pin reinforcement. In the second section of this chapter the impact behaviour of the pin reinforced sandwich is simulated and validated by experimental results. Two modelling approach of the pin reinforced foam core were tested and evaluated. In both simulations an analysis approach based on the building block approach (BBA) has been used to build the final simulation models.

5.1 Fundamentals of explicit simulation

The Finite Element Method (FEM) is the most used computational tool to develop industrial products. It is used in many sectors, like automotive, aerospace, wind turbine as well as medical products, and can give solutions for different problems like crash, fluid dynamics and kinematics of mechanical systems. It enables engineers to find accurate solutions to problems that cannot be solved by using standard analytical methods.

Generally, the FEM cannot replace experimental investigations, as they are important to generate material properties and to validate the FEM results. In the aerospace industry the numerical simulation is performed parallel to the experimental investigations following the Building Block Approach (BBA) used to design and certify aerospace structures. Beginning from the coupon level the numerical models are validated, the complexity of the simulation and the tests is increased until reaching the full scale test level. The simulation helps engineers to understand the structural behaviour and the manufacturing processes and leads to the reduction of structure development time and testing costs. A validated simulation model could be used to further improvement of the structure or development of similar structures as long as the design principles are the same and the damage behaviour of the material is validated.

5.1.1 General aspects and basic equations

The FEM is a powerful tool to solve problems marked by irregular geometry and/or nonlinear material/process behaviour and unusual loading and/or boundary conditions. As the nature behaviour is usually governed by equations expressed in integral or differential form, the FEM gives solutions to these mathematical equations with improved accuracy compared to analytical methods, by discretisation a body or an environment into a finite number of elements and nodes, and transforming the governing expressions into algebraic equations. To solve a solid mechanic problem the requirements of equilibrium, compatibility, material constitutive equations and boundary conditions should be satisfied. The algebraic equations are solved in every node, taking into consideration the above mentioned requirements, and solutions are given in the form of a set of numbers that gives information about the stress, displacement and other states of the body. A standard FEM program consists of a pre-processor, a solver and a postprocessor. During pre-processing a mechanical problem is analysed, assumptions are made, and the loads and boundary conditions are defined. Then the mechanical model is discretised and meshed, the material models, the loads, the boundary conditions and other simulation parameters are applied. Afterwards, the finished model is sent to the solver to be calculated. After finishing the calculation the results are sent to the post-processor to be translated into a graphic representation showing the state of the structure. The user communicates with the FEM program using a graphic interface that enables the preparation of the model and the interpretation of the results. The same procedure is applied to other kind of numerical problems. The FEM calculation procedure is schematically presented in Figure 65.

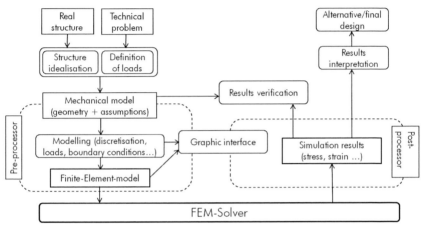

Figure 65. General procedure of FEM-analysis (adapted from [94])

Depending on the loading conditions different solvers are used. For linear and nonlinear static, linear dynamic and last but not least coupled temperature-displacement (quasi-static) implicit solver like Abaqus/Standard or Nastran are typically used. Explicit solvers like Abaqus/Explicit and LS-Dyna are generally used to simulate high-speed dynamics, for instance impact and crash analysis of structures, as well as large, nonlinear, quasi-static analysis. The base equation for static and dynamic problems is the same, but the integration scheme used to solve the differential equations in time domain is different. In this work the explicit solver Abaqus/explicit was used to simulate the low-velocity impact and the quasi-static tests. The general equilibrium equation for a discretised finite element model is:

$$M^{NM}\ddot{u}^M = P^N - I^N. \tag{5.1}$$

Where P^N is the external force vector, I^N is the internal force vector created by stresses in the element, $M^{NM}\ddot{u}^M$ is the force vector due to material inertia and the superscript N is the degree of freedom of the model. This equilibrium equation can be applied to any mechanical systems, considers all nonlinearities and can be solved incrementally using implicit or explicit methods. The implicit and the explicit method differ in the determination of the unknown values at the current time. While the unknown values are calculated from already known information when the explicit method is used, the unknown values at the current time during implicit calculation are obtained from current information.
The internal forces can be written as:

$$I^N = \int_V \beta^N : \sigma dV. \tag{5.2}$$

With V is the current volume of the model, $\sigma(x)$ is the stress at a point currently located at position x and $\beta^N(x)$ is the strain rate-displacement rate transformation defined as:

$$\dot{\varepsilon} = \beta^N \dot{u}^N. \tag{5.3}$$

When the d´Alembert forces $(M^{NM}\ddot{u}^M)$, containing the mass matrix M^{NM} and the acceleration vector $\ddot{u}^M$, are small enough or constant with time, the dynamic terms of the equilibrium equation are set to zero and the equation can be used to solve static problems. Abaqus/Standard uses Newton´s method to solve static problems with modest nonlinearity. The implicit procedure assumes that the model matrices, the external loads and the displacement $u_{(i)}^N$ at iteration (i) are known. This approach is very time consuming as reordering of the elements and inversion of the mass matrix are required at each and every iteration. Following the Newton´s method the estimation of the equilibrium equation at iteration time (i) using a Taylor series expansion is:

$$P^N - I^N + \left(\frac{\partial P^N}{\partial u^M} - \frac{\partial I^N}{\partial u^M}\right) c^M + \cdots = 0. \tag{5.4}$$

This equation can be written as:

$$K^{NM} c^M = P^N - I^N. \tag{5.5}$$

Where

$$K^{NM} = \frac{\partial P^N}{\partial u^M} - \frac{\partial I^N}{\partial u^M}. \tag{5.6}$$

is the Jacobean matrix and c^M is the correction to the solution at degree of freedom N.

The incremental displacement Δu at iteration $(i+1)$ can now be calculated by the following expression:

$$\Delta u_{(i+1)}^N = \Delta u_{(i)}^N + c_{(i)}^N. \tag{5.7}$$

Every iteration step is checked and repeated in each increment until convergence is achieved by fulfilling the contact conditions, the force and the moment equilibrium at every node and the displacement corrections are small compared to the correction to the solution c^M. If the model contains strong nonlinearities like complex contact conditions or material plasticity and high deformation rates, the implicit solution scheme can lead to non-convergence and a solution is in that case impossible.

In case of strong nonlinearities or high-speed dynamic problems, the explicit time integration is an effective method to solve the equation of motion. It uses small inexpensive time increments and neither iteration nor convergence checking is needed to calculate the unknown values. The state of dynamic equilibrium at the start of the current time increment (t) can be written as follow:

$$M^{NM} \ddot{u}^M|_t = (P^N - I^N)|_t. \tag{5.8}$$

From equation (5.8) the nodal acceleration can be calculated using the following equation:

$$\ddot{u}^M|_t = [M^{NM}]^{-1} \cdot (P^N - I^N)|_t. \tag{5.9}$$

The explicit procedure doesn´t require the time consuming mass matrix inversion as it uses a diagonal mass matrix for more efficiency.

Now the velocities and the displacement can be updated by using the central difference integration rule as follow:

$$\dot{u}^N|_{t+\frac{\Delta t}{2}} = \dot{u}^N|_{t-\frac{\Delta t}{2}} + (\frac{\Delta t|_{t+\Delta t}+\Delta t|_t}{2}) \cdot \ddot{u}^N|_t \qquad \text{and} \tag{5.10}$$

$$u^N|_{t+\Delta t} = u^N|_t + \Delta t|_{t+\Delta t} \cdot \dot{u}^N|_{t+\frac{\Delta t}{2}}. \tag{5.11}$$

Compared to the implicit method the explicit is very efficient as it doesn´t require any convergence checking or mass matrix inversion, but the stability of the central difference integration rule and the simulation efficiency depends strongly on the time increment Δt. A stable time increment is required to enable convergence of the simulation. The stable time increment is determined based on the Courant-Friedrichs-Lewy (CFL) condition. It specifies that the time increment should be smaller than the time required for a sound wave to pass through the smallest element in the model. In Abaqus/Explicit the stable time increment is defined as follow:

$$\Delta t = \min \left(\frac{L^e}{c_d}\right). \tag{5.12}$$

Here L^e is the characteristic length of the element and c_d is the dilatational wave speed of the material. For a linear elastic material c_d is:

$$c_d = \sqrt{\frac{\lambda+2\mu_l}{\rho}}. \tag{5.13}$$

Where λ and μ_l are Lame´s constant and ρ is the material density. The stable time increment primary depends on element length and material density. In simulations with lightweight materials and high mesh resolution the stable time increment is very small and can increase the simulation effort. Simulation programs offer a technique called mass scaling to artificially increase the density of elements with small stable time increments. This method should be used with precaution as it can alter the simulation results by extensive use.

The conditional stability of explicit integration methods requires a large number of time steps. If an efficient solution of the individual time steps is desirable, this can be achieved by using elements with reduced numerical integration. Figure 66 shows the difference between full and reduced numerical integration using a bilinear quadrilateral element.

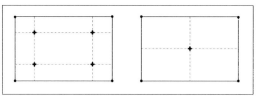

Figure 66. Numerical integration of a quadrilateral element using Gauss-quadrature. Left: full-integrated element; Right: Element with reduced integration [95]

In the FEM, the Gauss-quadrature is one of the most used integration method. The accuracy of the method depends on the number of integration points used (integration order). For an exact integration of the quadrilateral element at least 4 interpolation points are required. The 2x2 integration is therefore referred to as full integration [95]. In the case of reduced integration, the order of integration is reduced by one. For the element shown on the right side of Figure 66 one integration point remains in the middle of the element.

Hourglassing can occur when the stiffness matrix of the element with reduced integration is integrated. Hourglassing is defined as impermissible deformation of an element in which no strain is captured in the integration point. Since the strain energy becomes zero, these deformation states are also called zero-energy eigenmodes [95]. With disappearing strains the stresses disappear in the element too; accordingly the element does not resist the deformation. The hourglass mode can therefore propagate in the FE-mesh. In bilinear quadrilateral elements, contiguous elements form the shape of an hourglass, giving the name to the numerical phenomenon.

Hourglassing occurs because the numerical integration takes into account only the deformation of the integration points. Figure 67 provides an example for this. The already known bilinear rectangular element under moment loads is shown. When fully integrated, the nodes are shifted from the undeformed reference configuration. The element registers this as strain change. In the case of reduced integration, the integration point lies at the intersection of the neutral axes. By definition, there is no deformation in this point. Visually, this can be recognized by the fact that the position of the integration point remains unchanged.

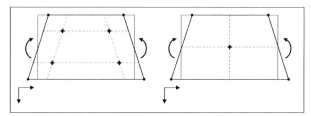

Figure 67. Left: element with full-integration (no hourglassing); Right: element with reduced integration (hourglassing) [95]

There are three approaches to avoid hourglassing [95]. The first possibility is to use quadratic approach functions. As the number of nodes increases, the number of hourglass modes decreases. In addition, the corresponding fashion cannot propagate over several elements. Second, mesh refinement can reduce hourglaassing effects. The higher number of nodes per area or volume reduces the likelihood of hourglassing. Third, a suitable stabilisation method can lead to a significant weakening of the hourglass problem. Under this condition, linear elements with reduced integration can still be used. Most of the commercial simulation programs provide different stabilisation methods to reduce hourglass effects. However, hourglass stabilisation introduces artificial energies into the system that affect the accuracy of the simulation results. The user must ensure that the artificial energy remains negligibly small compared to the total energy of the system.

5.1.2 Contact modelling with Abaqus

The simulation program Abaqus/Explicit offers a number of contact modelling options. But first, before setting the contact algorithm the mechanical contact properties should be defined. The Coulomb friction law is generally enough to model the contact properties. It defines the relationship between normal and frictional forces at the interface of two solids as follows:

$$F_R = \mu F_N. \tag{5.14}$$

The factor μ is called friction coefficient. It depends on the combination of materials, the surface condition and the relative speed of movement of the contact bodies to each other. For low-velocity impacts, a speed-independent coefficient of friction can be approximated. For the present material combination of steel and CFRP as well as contact between CFRP-plies a friction coefficient $\mu = 0.3$ is assumed. This coefficient of friction is applied to all simulations performed in this work.

After defining the contact properties the contact algorithm can be set. Abaqus/Explicit offers the possibility to choose between general contact formulation and contact pairs formulation. The general contact allows the modelling of various types of contact without need to define the respective contact type and the contact surfaces. The contact pairs and contact conditions are identified automatically and less contact restrictions of the types of surfaces involved are generated. The general contact algorithm is generally faster than the contact pair formulation and is efficient at pre-processing of models with multiple components and complex topology. As the name implies, in the contact pairs algorithm, every contact pair interaction should be clearly defined. Special attention should be paid to consider every possible contact. With large models, however, much longer calculation times than with the general contact formulation are the result. This contact formulation is suited for contact with special conditions that differs from the parameters of the general contact. Combining both contact formulations is possible in Abaqus/Explicit.

The general contact formulation uses the penalty method algorithm to enforce the contact constraint. With the penalty contact algorithm, also known as soft contact, the contact condition is not exactly met, as it allows a defined penetration of the slave node into the master surface. At each penetrating contact node, a spring is introduced which counteracts the penetration (see Figure 68) creating an additional artificial stiffness in the model, which would reduce the critical stable time increment. The spring stiffness determines the maximum penetration and the contact force, thus the accuracy of the contact modelling. Abaqus/Explicit chooses the spring stiffness automatically to minimise the effects on time increment. Nasdala [96] recommends the penalty method for dynamic-explicit simulations, since only small intersections can occur due to the small time increments, and consequently the quality of the results is only slightly influenced.

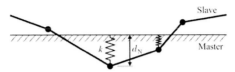

Figure 68. Schematic presentation of the penalty method with the introduced spring elements with k is the spring stiffness and d_N the slave node penetration [96]

The contact pair's formulation uses the kinematic contact algorithm by default. It enables a precise compliance with the contact conditions, as it allows no penetrations. The kinematic contact algorithm uses a predictor/corrector contact algorithm to predict the necessary contact force in the next step in order to avoid node penetration. Since no artificial energy is introduced to the system with the kinematic contact, the critical time increment is unaffected by this contact formulation.

5.1.3   Modelling the face sheet damage

The face sheets used in this work were made of carbon fibre triaxial non crimp fabric (NCF). Every NCF-layer consists of three UD-layers that have equal areal weight but different orientations. These UD-layers are stitched together with PES-yarn forming a semifinished preform. The textile manufacturing process creates fibre waviness and stitching yarn gaps, which affects the compressive properties of the UD-layer stronger than the tensile properties. Typical practice to simulate NCF-materials is to consider every single layer as UD-layer, properties are adopted from similar UD- or Prepreg-materials and the manufacturing effects are considered by knock-down factors for the stiffness and the strength.

Since there was no similar UD-material available, tensile tests were performed using specimens with different layups and the stiffnesses were determined. The UD-equivalent stiffnesses were obtained by analytical rules of mixture, while the strength and the interlaminar properties were obtained from the literature and calibrated by the simulation of three-points bending test.

For an accurate simulation of the damage behaviour of laminated CFRP the intralaminar and interlaminar damages should be considered. The intralaminar damages include the damage mechanisms within an individual layer in a laminate, like fibre rupture or matrix cracking under tension and shearing. The interlaminar damages describe the material failure occurrence between the individual plies, which is dominated by the resin system properties.

A failure criterion describing the damage initiation in a laminate is insufficient to describe the failure propagation after low-velocity impact. A model describing the damage propagation should be used to describe the complete failure process. As fibre-reinforced materials like CFRP have anisotropic elastic behaviour with brittle damage, Abaqus/Explicit assumes a bi-linear behaviour of the equivalent stress and displacement. The undamaged material behaves linear elastic until a damage initiation criterion is fulfilled, then the material begins to degrade linearly (negative slope) until a predefined energy G_c is dissipated and the failed element is deleted. This energy is a material parameter and represents the dissipated energy at defined failure mode. This bi-linear behaviour is applied to all failure modes. A representative evolution of stress and strain at failure is depicted in Figure 69.

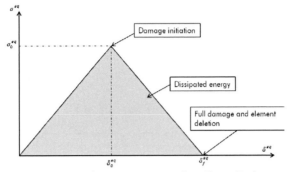

Figure 69. Linear damage evolution for composite materials in Abaqus/Explicit

For the onset of degradation in elements with plane stress formulation and composite materials properties, Abaqus/Explicit [97] uses a damage initiation criteria based on the phenomenological failure criterion Hashin´s theory [98, 99], which distinguish between four damage initiation mechanisms: fibre failure in tension and compression and matrix failure in tension and compression. The specific failure criteria are defined as follow:

Fibre tensile failure onset ($\bar{\sigma}_{11} \geq 0$):

$$F_f^t = (\frac{\bar{\sigma}_{11}}{X^T})^2 + \alpha_{HS}(\frac{\bar{\tau}_{12}}{S^L})^2. \tag{5.15}$$

Fibre compression failure onset ($\bar{\sigma}_{11} < 0$):

$$F_f^c = (\frac{\bar{\sigma}_{11}}{X^C})^2. \tag{5.16}$$

Matrix tensile failure onset ($\bar{\sigma}_{22} \geq 0$):

$$F_m^t = (\frac{\bar{\sigma}_{22}}{Y^T})^2 + (\frac{\bar{\tau}_{12}}{S^L})^2. \tag{5.17}$$

Matrix compression failure onset ($\bar{\sigma}_{22} \geq 0$):

$$F_m^c = (\frac{\bar{\sigma}_{22}}{2S^T})^2 + \left[(\frac{Y^C}{2S^T})^2 - 1\right]\frac{\bar{\sigma}_{22}}{Y^C} + (\frac{\bar{\tau}_{12}}{S^L})^2. \tag{5.18}$$

Where F_f^t, F_f^c, F_m^t and F_m^c are the damage onset factors of the fibre and matrix under tensile and compression loading; X^T and X^C are the longitudinal tension and compression strengths; Y^T and Y^C are the transverse tension and compression strengths. S^L and S^T are the longuitudinal and transverse shear strengths; α_{HS} is the contribution factor of the shear stress to the fibre tensile failure onset; and $\bar{\sigma}_{11}$, $\bar{\sigma}_{22}$ and $\bar{\tau}_{12}$ are the components of the effective stress tensor $\bar{\sigma}$ that considers the actual damage situation and is calculated from:

$$\bar{\sigma} = M\sigma. \tag{5.19}$$

Where σ is the true stress and M is the damage operator, which is defined as follow:

$$M = \begin{pmatrix} \frac{1}{(1-d_f)} & 0 & 0 \\ 0 & \frac{1}{(1-d_m)} & 0 \\ 0 & 0 & \frac{1}{(1-d_s)} \end{pmatrix}. \tag{5.20}$$

Where d_f, d_m and d_s are the internal damage variables for fibre, matrix and shear damage respectively. The shear damage variable d_s is calculated from the damage variables for fibre and matrix.

After damage onset the degradation of the material starts, its response is controlled by the fracture toughness of each failure mode and it is computed from:

$$\sigma = C_d \varepsilon. \tag{5.21}$$

Where ε is the strain and C_d is the damaged elasticity matrix that considers the stiffness degradation after damage onset and it is defined as follow:

$$C_d = \begin{pmatrix} (1-d_f)E_1 & (1-d_f)(1-d_m)\vartheta_{21}E_1 & 0 \\ (1-d_f)(1-d_m)\vartheta_{12}E_1 & (1-d_m)E_2 & 0 \\ 0 & 0 & (1-d_s)GD \end{pmatrix}. \tag{5.22}$$

Where $D = 1 - (1-d_f)(1-d_m)\vartheta_{12}\vartheta_{21}$. d_f, d_m and d_s are the variable describing the current damage state for the fibre, matrix and shear damages respectively. E_1, E_2, G, ϑ_{12} and ϑ_{21} are the Young´s Modulus in the fibre direction, the Young´s modulus in the matrix direction, the shear modulus and the Poisson´s ratios respectively.

As the fracture energy at a given mode G_{ce} is mesh dependent and is proportional to the volume of the element, a characteristic length L_c is introduced to define the damage evolution through equivalent displacement rather than strain. This manipulation moderates the mesh dependency [100] and avoids zero fracture energy for fine meshes. But, following the equation (5.23) proposed by Lapczyk and Hurtado [101] keeping the fracture energy G_{ce} constant alters the fail strain of the corresponding damage mode.

$$\varepsilon^f = \frac{2G_{ce}}{\sigma_c L_c}. \tag{5.23}$$

The experimental and theoretical background of the fracture energies associated to every damage modes is provided by Maimí et al. [102]. However, the proposed tests to determine the fracture toughness of fibre damage under tensile and compression loads are not carried out following a standard specification and these parameters are rarely available in the open literature, so that assumptions should be made and calibrations by the simulation should be performed.

The interlaminar interface of a laminated CFRP is mainly composed of a resin rich layer, which dominates the mechanical properties of the interface. The main failure mode of the interface is delaminations, which consists of cracks that emerge due to normal loading or sliding, and then propagate between the laminate plies leading to degradation of laminate strengths. Three basic modes of crack tip deformation can be distinguished (see Figure 70): Tension opening (Mode I), in-plane shear (Mode II) and out-of-plane shear (Mode III). Only mode I and mode II are applicable to laminated composite materials.

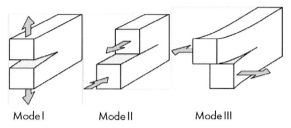

Mode I   Mode II   Mode III

Figure 70. Opening modes of crack tip deformation [103]

In the last decade the cohesive zone model (CZM) has been extensively used to simulate delamination after low-velocity impact [104-107]. Due to its efficiency, this model was integrated to various simulation programs, like Abaqus and LS-DYNA, by providing special elements, based on the work of Camanho and Davila [108], called cohesive zone elements. The CZM applies the linear elastic fracture mechanic (LEFM) to simulate the crack opening between two adhered surfaces. It can be applied to model delamination between two laminate layers as well as decohesion between two bonded surfaces. Compared to other crack prediction techniques like the virtual crack closure technique (VCCT) the CZM offers many advantages like computational efficiency, as the CZM-mesh is generally coarser than the VCCT-mesh, and no initial crack is necessary to model crack propagation, since cohesive elements are applied in potential delamination areas (interface between unidirectional plies) and cracks initiate when specific conditions are fulfilled.

The CZM assumes a linear elastic behaviour prior to damage onset based on traction-separation law. When a stress or strain condition is reached the stiffness of the cohesive element begins to degrade progressively until the dissipated energy calculated from the traction-separation diagram is equal to the critical fracture energy of the interface, and the element is then deleted. The global behaviour is similar to the damage behaviour of laminated CFRP illustrated in Figure 69. In this way, delamination damage is incorporated into the cohesive elements and no initial crack is required, since the crack may initiate and propagate anywhere in the cohesive element layers. However, in order to model the delamination accurately at least three elements are required in the cohesive zone, which may lead to mesh refinement and to the increase of computational time. The cohesive zone is defined as the area with material softening at the crack tip (see Figure 71) [109]. Song et al. [110] proposed the following equations to calculate an approximate cohesive zone length for mode I and mode II:

$$l_{cz-I} = M_{cz}E_{czm}\frac{G_{Ic}}{T_0^2} \qquad \text{and} \qquad (5.24)$$

$$l_{cz-II} = M_{cz}E_{czm}\frac{G_{IIc}}{S_0^2}. \qquad (5.25)$$

The variable M_{cz} depends on the used cohesive zone model and lies between 0.21 and 1. E_{czm} is the cohesive zone stiffness, T_0 corresponds to the resin tensile strength, S_0 is the shear strength of the resin system, G_{Ic} and G_{IIc} are the critical energy release rates under mode I and mode II respectively. As at least three cohesive elements within the cohesive zone are suggested, the following equation should be used to verify this condition:

$$l_e \leq \frac{l_{cz-I}}{N_e} \qquad \text{and} \qquad (5.26)$$

$$l_e \leq \frac{l_{cz-II}}{N_e}. \qquad (5.27)$$

Where l_e and N_e are the length of the cohesive element and the number of cohesive elements within a cohesive zone respectively.

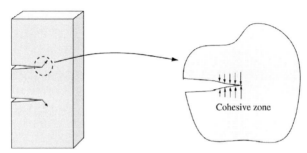

Figure 71. Schematic presentation of the cohesive zone [111]

The CZM available in Abaqus assumes a linear elastic behaviour until the damage onset. The traction stress vector t is related to the separation δ by mean of an elastic constitutive matrix. The traction and separation vectors have three components, which represents the normal (n) and the two shear directions $(s$ and $t)$. The separation is defined as the change in displacement between the top and bottom nodes of the cohesive element. The dissipated energy is calculated from the area under the traction-separation curve.

In order to start the softening of the interface (degradation) a damage initiation criterion should be specified. Abaqus [112] gives the possibility to choose between four different damage onset criteria: maximum nominal stress criterion, maximum nominal strain criterion, quadratic nominal stress criterion and quadratic nominal strain criterion. The first two criteria assume that damage initiates when the maximum nominal strain or stress ratio in normal or shear direction reaches a value of one. That means there is no interaction between the different damage modes. The last two criteria assume a quadratic interaction function between the nominal stress/strain ratios. When the damage onset expression reaches a value of one the degradation starts. The quadratic nominal stress criterion was chosen in this work and is expressed as follow:

$$\{\frac{\langle t_n \rangle}{t_n^0}\}^2 + \{\frac{t_s}{t_s^0}\}^2 + \{\frac{t_t}{t_t^0}\}^2 = 1 . \qquad (5.28)$$

Where t_n^0, t_s^0 and t_t^0 are the nominal stresses for purely normal deformation or purely shear deformation in first or second shear direction respectively. For laminated composites t_s^0 and t_t^0 are generally assumed to be equal to the shear strength of the resin system.
The used Macaulay brackets mean that pure compressive stresses or deformations do not lead to cohesive damages.

After damage initiation the material stiffness is degraded and the occurred damages are not any more reversible. A scalar variable, D ranging from zero at damage initiation to one at cohesive failure, is introduced to represent the overall damage in the material. It considers combined effects of the damage mechanisms. The stress components of the cohesive zone elements after damage initiation are then calculated as follow:

$$t_n = \begin{cases} (1-D)\bar{t}_n, & \bar{t}_n \geq 0 \\ \bar{t}_n, & \text{otherweise (no damage at compression)} \end{cases}, \qquad (5.29)$$

$$t_s = (1-D)\,\bar{t}_s\,, \qquad (5.30)$$

$$t_t = (1-D)\,\bar{t}_t\,. \qquad (5.31)$$

Here $\bar{t}_n$, $\bar{t}_s$ and $\bar{t}_t$ are the predicted stress components by using the linear elastic traction-separation low for the current strain without damage.

An effective mixed-mode displacement δ_m [108] is introduced to consider the combination of normal and shear deformation across the interface and is defined as:

$$\delta_m = \sqrt{\langle \delta_n \rangle^2 + \delta_s^2 + \delta_t^2}\,. \qquad (5.32)$$

Abaqus gives the possibility to choose between linear or exponential softening as well as defining D in tabullar form between damage onset and final failure. For the linear stiffness degradtion used in this work the damage variable D is expressed as follow:

$$D = \frac{\delta_m^f (\delta_m^{max} - \delta_m^0)}{\delta_m^{max}(\delta_m^f - \delta_m^0)}\,. \qquad (5.33)$$

In equation 5.33 δ_m^0 refers to the effective displacement at damage initiation, δ_m^f is the effective displacement at complete failure and δ_m^{max} is the maximum value of effective displacement achieved during load history.

An illustration of the mixed-mode response in cohesive elements is given in Figure 72. For laminated composites the mixed-mode interaction can be considered by the definition of a Benzeggagh-Kenane parameter η_{BK}, which is a material property.

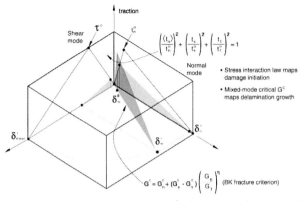

Figure 72. Mixed-mode Traction-separation response of cohesive zone elements [112]

While the critical energy release rates for mode I and mode II damages and their interaction can be derived experimentally, by means of Double Cantilever Beam (DCB), End Notched Flexure (ENF) and Mixed Mode Bending (MMB), the stiffness parameters of the cohesive zone are not well defined. Many suggestions [109, 113-115] concerning the calculation of the cohesive zone stiffness K_{CZ}, also called penalty stiffness, exist in the open literature and lead generally to good results. It is generally suggested that $K_{CZ} = K_n = K_s = K_t$. While Camanho et al. [113] used a penalty stiffness of $K_{CZ} = 10^6 \ N/mm^3$, Turon et al. [109] suggested the following equation to calculate K_{CZ}:

$$K_{cz} = \frac{\alpha_{cz} E_{33}}{h_{cz}}. \qquad (5.34)$$

Where E_{33} is the E-modulus in thickness direction of the laminate, h_{cz} is the thickness of the sublaminates (bonded plies) and α_{cz} is a parameter between 1 and 50 to adjust the stiffness of the cohesive zone elements, so that the penalty stiffness is large enough to not increase the compliance of the model without inducing numerical instabilities.

As previously mentioned, at least three cohesive zone elements are necessary in the cohesive zone for a proper delamination modelling. High strength values of the resin system can lead to short cohesive zones and thus increase the computation time. Turon et al. [109] suggested in his work based on the research of Alfano and Crisfield [116] an engineering solution to enable the use of coarse mesh when modelling delamination growth. Artificially reducing the interface strength while maintaining the fracture energies (G_{Ic} and G_{IIc}) constant, would allow the use of coarser mesh and thus reducing the computational time without changing the material behaviour. Turon proposed the following expressions to calculate the artificial strength of the interface, where the length of the element l_e, the number of elements in the cohesive zone N_e, the cohesive zone stiffness E_{czm} and the fracture toughness are known and used as inputs:

$$T^a = \sqrt{\frac{M_{cz} E_{czm} G_{IC}}{N_e l_e}}. \qquad (5.35)$$

$$S^a = \sqrt{\frac{M_{cz}E_{czm}G_{IIC}}{N_e l_e}}. \tag{5.36}$$

The cohesive zone strengths used in the simulation should be the minimum of T^a and T_0, and of S^a and S_0. This approach is effective to reduce computational time but at expense of accuracy by predicting stress concentrations at crack tips.

5.1.4 Modelling the crushing behaviour of the foam core

Closed cell foams have generally an initial linear elastic behaviour until compression failure. Due to the small value of compressive strength, the linear region is very small and irreversible deformations start already at very small deflections of the face sheet during impact loading. The post failure behaviour is characterised by crushing of the foam at nearly a constant load level. Core crushing is the main energy consuming mechanism and it is described as cell wall collapse, as a combination of crushing/buckling and the formation of plastic hinges in the cell walls. The density of the foam starts to increase slowly until the full compaction and lock-up of the cell walls where the stress in the core increases sharply.

To model the irreversible deformation of the foam core the crushable foam model implemented in Abaqus/Explicit was chosen. Since the tensile strength of the used PMI foam differs from its compressive strength the phenomenological constitutive model "the volumetric hardening model" was used to simulate the compressive response of the foam during the crushing process. On the one hand, the model assumes that the foam cell deformation under compression is plastic for short duration events. On the other hand, the tensile behaviour of the foam is assumed to be brittle and the load bearing capacities are smaller than under compressive loads. For pure shear and negative hydrostatic pressure stress state a perfect plastic behaviour of the foam is assumed. Finally, the model assumes foam hardening for positive hydrostatic pressure stress states. This model can only be used with the linear elastic material model. Predicting shear crack initiation and evolution is not possible with this model, as no damage initiation criterion and damage evolution law are implemented.

The yield surface of the volumetric hardening model is assumed an ellipse in the meridional (p-q) stress plane. It depends on both deviatoric and hydrostatic stresses q and p that are defined as:

$$q = \sqrt{\frac{3}{2}S{:}S}. \tag{5.37}$$

$$p = -\frac{1}{3}trace(\sigma). \tag{5.38}$$

Where p, q and S are the pressure stress, the Mises stress and the deviatoric stress respectively. To proper build the yield surface the uniaxial compression yield stress as function of the corresponding plastic strain and the two strength ratios, i.e. the compression yield stress ratio k and the hydrostatic yield stress ratio k_t, which are the ratio of the uniaxial compressive strength ($\bar{\sigma}_{cz}$) to the hydrostatic compressive strength (p_c^0) and the ratio of hydrostatic tensile strength (p_t) to the hydrostatic compressive strength respectively.

The volumetric hardening model assumes that the point on the yield ellipse in the meridional plane representing the hydrostatic tension loading is fixed and the evolution of the yield surface is controlled by the volumetric compacting plastic strain experienced by the material. As p_t remains constant, the evolution of the yield surface can be expressed through the evolution of the yield stress in hydrostatic compression p_c, which can be expressed as:

$$p_c(\varepsilon_{pl}^{vol}) = \frac{\sigma_c(\varepsilon_{pl}^{vol})\left[\sigma_c(\varepsilon_{pl}^{vol})\left(\frac{1}{\alpha^2}+\frac{1}{9}\right)+\frac{p_t}{3}\right]}{p_t+\frac{\sigma_c(\varepsilon_{pl}^{vol})}{3}}.$$ (5.39)

Where ε_{pl}^{vol} the volumetric plastic strain, which is equal to the uniaxial compression strain from the uniaxial compression test. σ_c, p_t and α are the uniaxial compressive stress, the hydrostatic tensile strength and the shape factor respectively. The shape factor α can be computed using the compression yield stress ratio k and the hydrostatic yield stress ratio k_t. The yield surface of the crushable foam core with volumetric hardening model is illustrated in Figure 73.

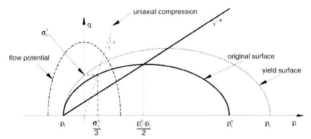

Figure 73. Yield surface of the volumetric hardening foam model [117]

5.2 Numerical simulation of impact on a curved foam core sandwich shell for aerospace application

This section demonstrates the use of finite element method to predict the structural and damage response of a large-scale aircraft sandwich panel under low-velocity impact loading. The aerodynamic shape of the foam core sandwich shell and the connection to the aircraft structure were considered during impact tests and simulation. Building of the large scale simulation model was based on the building block approach (BBA). It considers a separate validation of the different components of the sandwich (CFRP face sheets and PMI-foam core). The complexity of the impact model was stepwise increased. A model for a generic flat sandwich specimen was first developed and validated; then the low-velocity impact behaviour of the large curved sandwich panel was simulated using the parameters from the previous simulation level. The simulation models were validated by comparison of numerical results with experimental results from impact tests performed in this work. In order to incorporate the face sheet delamination and interface debonding with minor increase in the computational costs, multi-scale modelling techniques were used.

The aim of this simulation was to collect know-how about simulating the impact behaviour of foam core sandwich structure as well as to test and validate the BBA-based modelling approach and the multi-scale modelling techniques. This step is essential to verify the impact model parameters before starting the validation of the pin reinforced foam core model due to the fact that less uncertainties and extensive model modifications are existing when adjusting the material and model parameters.

5.2.1 Literature review

Using finite element analysis in early development stages of a project gives more freedom to study different designs, which reduces the time and costs during the test phase. When more accurate information about the impact behaviour of a sandwich structure or a parameter study is required, the finite element analysis is typically used. Ivañez et al. [118] studied the effect of impactor incidence angle on the impact response of honeycomb sandwich structure. No remarkable difference in the maximum impact force and absorbed energy was observed for impact angles below 10°. It was found that for higher impact angles (over 15°) the peak load decreases while the absorbed energy increases. In order to predict the impact response for impact angles higher than 15° and to overcome the problems of the experimental setup at high impact angles, the numerical simulation was used. An impact model was validated through comparison of main impact parameters with experimental results for impact angles smaller than 15°, than the model was used to investigate the impact behaviour at impact angles up to 50°. In another study He et al. [119] investigated the effects of impact energy, foam core thickness and impactor shape and size on the impact behaviour of foam core sandwich. An impact model, which doesn't consider the delamination and debonding of the face sheet, was also developed and validated. A reasonable agreement with the experimental results was reached. Feng and Aymerich [106] simulated the impact behaviour of PVC foam core sandwich. They used an energy-based continuum damage mechanics models for the prediction of intralaminar damage in the damaged face sheet. The delamination and debonding of the skins were considered by using cohesive zone elements. As a result, the accuracy of the simulation was considerably improved compared to other simulations without interlaminar damage. Most of the model validations reported in the literature were only carried out for the final impact model. The different material properties were taken whether from experiments or other literature. Block [7] and Reinhardt [120] used an approach following the building block approach, typically used for certification of aerospace components, to develop their final impact models. They validated the different components of the sandwich structures separately using quasi-static test results. Then they built an impact model with the validated parameters. This approach enables to understand the different interactions between the model parameters and delivers simulation results with good accuracy.

Among the studies concerning the simulation of impact behaviour of sandwich structures, most of them considered flat and small sandwich specimens with predefined boundary conditions. The assumed specimen size and boundary conditions are not typical for aerospace applications. The aim in this section is to develop an impact model for a large foam core sandwich panel that considers the curved shape of the structure, the connection with the rest of the structure and the transition between the monolithic and sandwich section. The development of the model is based on the building block approach, the complexity of the model is increased step-by-step and a validation is performed in every step. To increase the accuracy of the model a multi-scale approach is used in the impact region. The delamination and debonding of the faces are considered by using the cohesive zone elements. Low-velocity impact tests using drop-weight impact system and mobile impact gun were performed to obtain experimental results for impact model validation

5.2.2 Modelling approach

The mechanical behaviour of composite materials is unique and depends strongly on material combination (fibre and matrix system), manufacturing process and environmental conditions. In order to reduce risks and costs during product development while satisfying the customer and certification requirements the Building Block Approach (BBA) has been widely used in aerospace industry [121]. The BBA is a design methodology intensively used in aerospace sector and employs tests and analysis from coupons to structural elements to full-scale component/system to certify structures for aerospace applications (Figure 74-a). Intensive material investigations are usually performed at coupon and element level to understand the material behaviour in critical conditions. The complexity and test costs increase when moving to the next level in the test pyramid but the number of test specimens decreases. Extensive testing of generic specimens leads generally to reduction of costs, specimen count and risks at higher testing level.

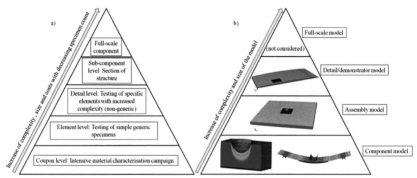

Figure 74. a) BBA for composite aerospace structure; b) Applied approach for impact simulation

An approach (Figure 74-b) based on the BBA (Figure 74-b) was applied for the impact simulation of foam core sandwich. This simulation approach is typically used to design aerospace structures. The two components of the sandwich panel, namely the face sheet and the foam, were separately validated by simulation of quasi-static tests and comparison with experimental results. Validation of the single components enables the adjustment of material parameters and better understanding of material models. The next step is to simulate and validate the low-velocity impact behaviour of generic sandwich panels. At this level, interactions with boundary conditions are controlled and the impact tests are repeatable, as a drop weight impact system was used. Hence, elimination of errors and repair of the simulation model can be done with reduced effort. If adjusting of material properties is necessary, the model of the face sheet and the foam core should be validated again. After validating all impact parameters for generic sandwich specimen, a sub-component model of the sandwich shell was developed using the knowledge gathered in the previous modelling level. The sub-component model considers the aerodynamic curvature of the top face sheet, realistic boundary conditions and the transition between the monolithic run outs and the sandwich section. Impact tests on a large curved sandwich panel with simulated realistic boundary conditions were performed using a mobile impact gun. Following this modelling strategy would lead to a fast validation of the sub-section model and usually without need to change any parameters already set in the generic impact model. Validation of a full-scale model was not performed in this work, as more complex and more expensive tests are needed and the impact behaviour of the sandwich panel would not change in a full scale impact test. The developed model was validated for the damage mode local face sheet damage with core crushing. The global shear failure mode was not observed during the impact tests, that´s why it was not possible to validate the model for this failure mode.

The sandwich panels tested in this section are made up of two thin carbon-epoxy face sheets separated by a low density PMI-foam core and were manufactured using the vacuum assisted resin infusion process as described in section 3.1. The face skins consisted of four layers Toho Tenax HTS40 carbon fibre Non-Crimp Fabrics (NCF) [65] impregnated by Hexcel RTM6 epoxy resin [64], which resulted in a face sheet thickness of about 1.5 mm and the layup $[(45°/0°/-45°)s]_2$. The closed cell PMI foam ROHACELL® 71 HERO [24] with a density of about 75 kg/m^3 was used as core material, and was milled to give the aerodynamic shape of the large sandwich panel.

5.2.3 Strain rate effects of the sandwich components

Compared to the behaviour during static loading, materials behave differently when subjected to short time dynamic loading with high strain rates. The materials may react with dynamic stiffening, strengthening and reduction of failure strain. This strain rate sensitivity depends on the material morphology, loading type and environment conditions. For composite materials this behaviour is very complex, as it depends on the properties of the used materials consisting of matrix system and reinforcing fibre. A wide range of scientific articles describe the strain rate sensitivity of composite materials and show the strong dependency on tested materials and strain rates. Depending on the strain rate range two testing methods are used [122]: for low and medium strain rates up to about 10 s^{-1} servo-hydraulic testing apparatus are typically used, while the Spit Hopkinson pressure bar (SHPB) is widely used to test material behaviour under high strain rate loading. Figure 75 shows a schematic diagram of the SHPB test setup. The main components of SHPB test setup are a striker bar, incident bar, transmission bar, absorption bar and the data acquisition system. The test specimen is placed between the incident bar and the transmission bar. The striker is propelled usually by pressurised nitrogen and impinges the incident bar. When the impact between the striker and the incident bar is initiated an approximately constant compressive stress pulse is created and it hits the test specimen. A part of the impulse is reflected to the incident bar and the rest is transmitted to the transmission bar. The reflection and transmission waves are measured by two strain gages and the signals are treated by the data acquisition system. The same test setup has been adopted with some modifications to take into consideration other loading conditions, for instance tensile and torsion loading.

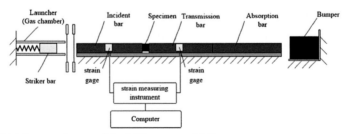

Figure 75. Schematic diagram of the SHPB test setup [122]

The most known technical fibres, namely carbon, glass and aramid show different behaviours when subjected to high strain rate loading. Figure 76 shows the ultimate tensile strength of T-700 carbon, E-glass, Kevlar and SiC fibres in function of the logarithmic strain rate. While the carbon fibre shows a strain rate insensitive behaviour, exhibit the other fibres a high sensitivity to the strain rate. The ultimate tensile strength of Kevlar-49 fibre increases linear to the logarithmic strain rate, the E-glass and SiC fibres show exponential increase, at small and medium strain rates these fibres are nearly strain rate insensitive then the ultimate tensile strength rises sharply for high strain rates.

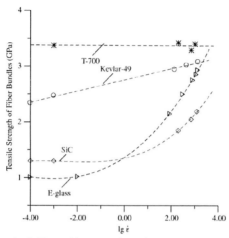

Figure 76. Tensile strength of different fibre types depending on the strain rate [123]

Zhou et al. [123, 124] investigated the tensile behaviour of T700 and M40J carbon fibre bundles under different strain rates. They concluded that the strain effect on the ultimate tensile strength and failure strain is negligible and carbon fibre can be considered as strain insensitive.

While the fibres show material depending strain rate sensitivity, exhibit thermosetting resins similar tendencies when tested at impact rates of strain. Thermosetting resins show generally a strain rate sensitive behaviour that depends strongly on the loading conditions. Gerlach et al. [125] performed compression, tension and 3-point bending at different strain rates using the thermosetting resin system Hexcel RTM-6. The compression and the tension tests showed for both tests an increase of the yield stress and initial stiffness with the increase of the strain rate. While the failure strain increased during compression test, an embrittlement and decrease of the strain to failure with increasing strain rate was observed during tension test. The 3-point bending test showed a good correlation with the tension test as the stiffness of the beam slightly increased and the ultimate deflection was extreme low at high strain rate compared to the quasi-static tests. Similar trends at high strain rates loading were observed by Buckley et al. [126] who tested three different thermosetting resins at impact strain rate under compression and tension. Gilat et al. [127] investigated the mechanical response of E-862 and PR-520 (toughened resin) resins in tensile and shear loading. Similar behaviour to the previous observation at tensile loading were reported, while the response in shear was ductile with an increasing maximum stress with increasing strain rate.

Depending on the material combination, the laminate layup and the loading direction the strain rate sensitivity may differ, which makes the characterisation of the strain rate dependency of composite material time and cost consuming. While the strain to failure under compressive load of the neat epoxy resin increases with increasing strain rate, the failure strain of CFRP and GFRP decreases with increasing strain rate [128]. Shokrieh and Omidi [129] investigated the strain rate dependent behaviour of unidirectional glass/epoxy composites tested in 0°-direction. The tensile strength, the tensile modulus and the absorbed energy showed a strain rate insensitive behaviour until about 2 s^{-1}, and then these properties increased sharply with the increase of the loading strain rate. The failure strain under tension remained constant until a strain rate of about 10 s^{-1}, and then it started to increase sharply with increasing strain rate. Similar tensile behaviour of GFRP was reported by Gurusideswar et al. [130]. Carbon fibre composites show totally different tensile behaviour. Taniguchi et al [131] determined the tensile properties of unidirectional CFRP in longitudinal and transverse direction as well as the shear properties under high strain rate up to 100 s^{-1}. On the one hand, the CFRP specimens showed no strain rate sensitivity in fibre longitudinal direction under tensile loads, the tensile strength and modulus remain constant in the investigated strain rate range. On the other hand, the tensile properties in transverse direction and the shear properties exhibit an increase with increasing strain rate. The shear modulus and strength showed a dramatic increase with the strain rate so that the shear strength exceeded the transverse tensile strength at strain rates over 50 s^{-1}. The dependence of these properties on the strain rate is depicted in Figure 77 . Comparing the dynamic and the static failure strains in Figure 78 showed that the tensile strain at failure in longitudinal direction remains constant (no sensitivity), increases with increasing strain rate when tested in transverse direction and the shear failure strain decreases under dynamic loads. Moreover, Taniguchi et al investigated the fracture surface using a scanning electron microscope (SEM) and they concluded that there is no difference between the static and the high strain damage behaviour in the 90°-specimens as the cracks propagate in both cases along the interface between the fibre and matrix. However, the fracture surface changes dramatically with increased strain rates in the [±45°]-specimens, as additional matrix cracks were observed in addition to the fibre/matrix interface damage observed in the static test. In addition, off-axis tensile tests under high strain rate were performed; the authors concluded that higher fibre orientation increase the strain rate sensitivity since the tensile behaviour is driven by the resin contribution. These conclusions concerning the dynamic tensile behaviour of unidirectional CFRP confirm the results presented by Gilat et al. who performed different dynamic tests at strain rates up to 600 s^{-1} [132].

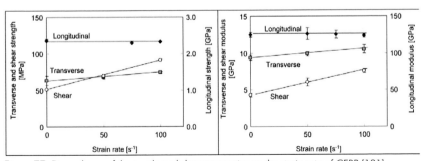

Figure 77. Dependence of the tensile and shear properties on the strain rate of CFRP [131]

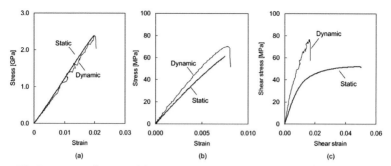

Figure 78. Comparison of static and dynamic stress-strain curves of CFRP: a) longitudinal direction; b) transverse direction; c) shear direction [131]

The strain rate effect on unidirectional carbon-epoxy specimens under longitudinal compression have been studied by Koerber and Camanho [133]. They developed a dynamic compression fixture for a SHPB to test flat rectangular specimens. They compared the SHPB results at an average strain rate of about 100 s^{-1} to quasi-static test results. They concluded that the compressive modulus in fibre direction is not strain rate sensitive up to a strain rate of 118 s^{-1}, while the longitudinal compressive strength drastically increased by about 40% with increasing strain rate. These conclusions are a validation of the results of Hsiao and Daniel [134] who observed only a slight increase of the dynamic compression modulus in fibre direction over the static value. Moreover, they found out that the strain at failure increases as the strain rate increases. Considering the transverse compression properties of UD epoxy carbon composites a moderate increase of the transverse compression modulus and a remarkable increase of the compressive strength with increasing strain rate are reported, while the ultimate compressive failure strain in transverse direction has no significant strain rate sensitivity [133, 134]. The strain rate sensitivity of carbon-epoxy laminates is a combination of all these observations reported in this section and it depends on the laminate layup. An accurate prediction of the strain sensitivity of laminate is only possible with extensive material testing.

Cellular polymeric foams exist with a wide range of mechanical properties. Depending on the parent material, the cell topology and the foam density, the material properties and the damage behaviour could be shaped to the aimed application. Arezoo et al. [135] studied the mechanical response of Rohacell PMI foams at different density under compression, tension shear and indentation. They performed in situ SEM during static compression and tension. They concluded that the damage behaviour of low-density PMI foam is driven by elastic buckling of the cell walls under macroscopic compressive loading, while the foams of high density failed due to plastic cell-wall bending. This strong dependency on cell topology makes the damage behaviour of closed cell foams under high strain rates unique and requires a thorough investigation of every single foam type. Williams and Lopez-Anido [136] investigated the strain rates and temperature effects on PVC and SAN foam core material under wave slamming condition ($\dot{\varepsilon}=2$ s^{-1}) and shear loading. They concluded that for both materials the shear modulus and shear strength increase with increasing strain rate compared to the quasi-static properties with a greater effect of strain rate on shear strength than on shear stiffness. These properties showed an inverse relationship with temperature. Many researches have been done to investigate the compressive response of PVC foam materials under high strain rate loading [137-140]. The following conclusions can be retained: The strain rate sensitivity and damage behaviour depends strongly on foam density, foam microstructure and strain rate as well as foaming direction. While the compressive stiffness has no significant strain rate sensitivity for all strain rate ranges, the compressive strength shows higher strain rate sensitivity. At low strain rates the compressive strength has a minor increase with increasing strain rate, but when the strain rate changes to higher level the compressive strength increases sharply, then the change within a strain rate range remains marginal. This behaviour is clearly illustrated in Figure 79 representing the strain-strain graphs of the PVC-foam HP200 with the density 200 g/m³ at different strain rates.

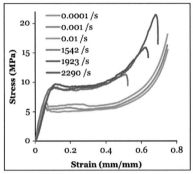

Figure 79. Stress-strain graphs of HP200 at different strain rates [140]

Concerning the influence of the foam density on the strain rate sensitivity, low-density PVC-foams seem to be less strain sensitive than PVC-foams of high densities. The compressive strength and the absorbed energy of high-density PVC-foams increase faster with the increase of the strain rate compared to the behaviour of low-density PVC-foams. The compressive strength and the plateau stress over the logarithmic strain rate of PVC-foams with different densities are depicted in Figure 80. The nominal densities are 60, 100, 200 and 250 kg/m³, which can be read in their nomenclature (HP60, HP100, HP200, and HP250). The absorbed energy can be associated with the plateau stress.

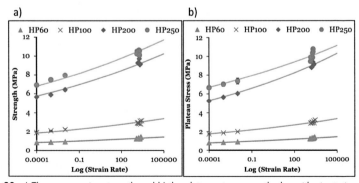

Figure 80. a) The compressive strength and b) the plateau stress over the logarithmic strain rate for PVC-foams [140]

Experimental investigations of the compression performance of Rohacell PMI foams were carried out in [63] and [91]. It was reported that the strain rate sensitivity is insensitive to foaming direction and depends strongly on the density of the PMI-foam. At low densities only minor effect on compressive strength was observed, while the stiffness increases with increasing strain rates. At high densities these tendencies are reversed, as the stiffness is nearly strain rate insensitive and the compressive strength increases with the increase of the strain rate. A mild strain sensitivity of PMI-foams is generally reported, the damage is ductile at low and medium strain rates, but the materials shows progressive embrittlement for strain rates beyond 100 s^{-1}. Figure 81 shows the strain rate sensitivity of the compressive strength of four different PMI-foams. Foams A, B, C and D have the densities 56, 71, 103 and 198 kg/m^3 respectively. On the one hand, the foams A and B show mild strain rate sensitivity, while foam C exhibit rate insensitive behaviour for the compressive strength. On the other hand, the foam D with nearly 200 kg/m^3 density shows a strain rate sensitive behaviour but with irregular tendency.

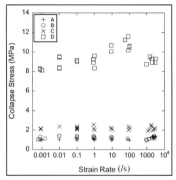

Figure 81. Sensitivity of the compressive strength of Rohacell foams to the strain rate [63]

Generally, the rate sensitivity of foam materials is strong dependent on the parent material, the foam density and the cell geometry. Most of the open source literature concentrated on the strain rate sensitivity of PVC and PMI foams under compressive loads. It can be concluded that low-density foams have reduced strain rate sensitivity compared to foams with high density. Moreover, a brittle damage behaviour at high strain rates is common for foam materials. While the compressive strength of PVC-foams shows a strain rate sensitivity that increases with increasing strain rate and foam density, the compressive strength of Rohacell foams shows a strain rate insensitive behaviour for low-density foams and an irregular sensitivity for foams with high density.

From the literature review dealing with strain rate sensitivity of foam cores and facings materials presented in this section, it is clear that dynamic loads with subsequent high strain rates have crucial effects on the mechanical behaviour of sandwich structures. However, considering strain rate sensitivity in low-velocity impact simulation using finite element method is not common practice, as neglecting them has only minor effects on the prediction accuracy like reported in many researches [7, 104, 106, 118].

In this work, it was decided to neglect the strain rate sensitivity of the used composite materials in the impact simulation for the following reasons: firstly, performing dynamic tests to investigate the strain rate sensitivity of the used materials would increase the experimental effort. Second, the used explicit solver Abaqus doesn't include material models that consider strain rate effects. Finally, impact simulations using similar materials and neglecting strain rate effects showed high accurate results. In order to verify the assumption of strain rate insensitivity of the used materials, it was decided to impact three sandwich panels with the same impact energy but with different impact velocity. The panels had a Rohacell HERO71 core with a thickness of 16.3 mm and 1.5 mm thick face sheet. The test setup and the materials were exactly the same as described in section 4.2. The impact energy was 35 J and the applied impact velocities were set to 3.7 m/s, 4.77 m/s and 5.88 m/s. This resulted in strain rates of about 190 s^{-1}, 244 s^{-1} and 301.5 s^{-1} respectively. Then the resulted damages and contact force-displacement diagrams were compared.

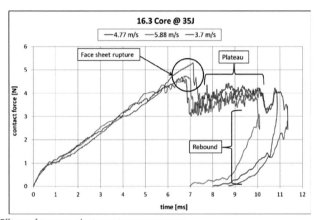

Figure 82. Effects of impact velocity on impact response

The generated contact force-displacement diagrams of the impact tests are depicted in Figure 82. No strain rates effects can be observed in the region until force peak (face sheet rupture) and in the plateau region. Only minor deviation are existent in the rebound region of the diagram (region after plateau) due to material stiffening. Moreover, the damage behaviour, composed of face sheet rupture with debonding and core crushing, was similar in the tested panels. The dent depth was between 2.26 mm and 2.65 mm. These results show that the tested sandwich panels have a non-significant strain rate sensitivity in the tested impact velocity range and validate the assumptions used in the impact simulation.

5.2.4 Face sheet model validation

With the aim to develop a reliable prediction of the face sheet damage propagation, the face sheet model validation was performed by means of simulation of quasi-static three-point-bending (3PB) test of the face sheet. The 3PB-Tests were performed by Block and are described in his dissertation [7]. The same material with the same stacking sequence as used in the impact specimens was tested in the global 0° and 90° directions according to DIN EN ISO 14125 [141]. Current practice to determine unidirectional (UD) layer properties (strength and stiffness) is to perform tensile tests in 0°, 90° and 45° using specimen made of UD layers. Applying this to NCF material is not possible, as the orientation of the UD layers could not be changed due to the stitching and the fixed orientation. In this work the stiffness properties of the used NCF-material were obtained as follow: standard tension tests following DIN EN ISO 527-4 [142] were performed using specimens with different layups. The stiffnesses of the different layups were determined, and then they were converted to UD-ply properties using the laminate theory. The results and the details of the tension test are reported in the appendix A2. The average thickness of the tensile specimens delivers a thickness of the equivalent UD-ply of about 0.133 mm. The strength and fracture mechanics properties of the equivalent NCF UD-ply were adopted from [7]. The mechanical properties of the face sheet used in the simulation are shown in Table 4.

Table 4. Mechanical properties of the equivalent UD-ply

Mechanical properties	Values
Density [kg/m³]	1510
Elastic properties [GPa]	$E_1 = 119$; $E_2 = E_3 = 8.5$; $G_{12} = 4.2$; $G_{23} = 3.8$; $\vartheta_{12} = 0.263$ [-]; $\vartheta_{23} = 0.2$ [-]
Strength [MPa]	$X^T = 2300$; $X^C = 1400$; $Y^T = 85$; $Y^C = 200$; $S^L = 125$
Fracture energies [kJ/m²]	$G_X^T = 100$; $G_X^C = 90$; $G_Y^T = 2.1$; $G_Y^C = 7.5$

A 3PB-model was created in Abaqus CAE. The indenter and the supports were modelled as rigid bodies. A constant velocity of 40 mm/s, which is faster than in the quasi-static 3PB-Test, was applied to the indenter. This doesn't affect the results since strain rate effects are neglected and inertia effects are not yet large in the simulation model. The contact interaction between the indenter, the supports and the 3PB-test specimen was modelled with the general contact algorithm provided by the Abaqus package, a friction coefficient of 0.3 was assumed. Due to the complex damage behaviour of the composite laminate an accurate damage prediction, that considers both intralaminar fibre/matrix damage and interlaminar damage between the plies in form of delamination, should be employed in the simulation. A suited approach to consider the latter failure modes is the ply-by-ply modelling, which was used in this work. Every single ply is modelled with one shell element layer (S4R-elements, 0.8 mm length) that is connected to the next ply with cohesive elements. The behaviour of the CFRP specimen is considered linear elastic until the damage initiates and the laminate begins to degrade. To consider the intralaminar failure modes the standard material model in Abaqus/Explicit based on Hashin criterion [98] for failure initiation and fracture energies for damage propagation was used. The material model distinguishes between tensile and compressive failure modes in the fibre and the matrix. The cohesive elements are responsible for the delamination and their constitutive mechanical model is based on the cohesive zone model with bilinear traction-separation law. The failed cohesive elements are deleted, so that the mechanical connection between the affected plies is eliminated and the kinematic of delamination could be accurately described. The used cohesive parameters in this work are summarized in Table 5 and are obtained from [7]. The quadratic nominal stress criterion and the linear damage evolution were used for the interlaminar damage model. In order to reduce the computational time, the expensive ply-by-ply modelling was only used in the central section of the 3PB-test specimen under the indenter; the full laminate further outward was modelled with a single stacked shell element. Both sections were connected using the kinematic coupling method available in Abaqus/Explicit. The simulation model of the 3PB-test is depicted in Figure 83.

Table 5. Summary of the parameters used for the cohesive elements

K_n [MPa/mm]	K_t [MPa/mm]	G_{IC} [J/m²]	G_{IIC} [J/m²]	t_n^0 [MPa]	t_s^0 [MPa]
$7.0*10^4$	$3.2*10^4$	426	1500	33	42

Figure 83. Simulation model of the 3PB-test

A comparison of the experimental and simulation results of the 3PB-tests in the laminate 0°- and 90°-directions is shown in Figure 84. While the model slightly overestimates the stiffness in the initial linear region and the damage propagation region for the 3PB-Test of the specimen in the 0°-direction, the simulation shows an accurate prediction of the linear elastic behaviour and the damage evolution in the 90°-direction. During the simulation of the 3PB-test of the 0°-laminate the specimen failed when the first ply on the tensile loaded side of the specimen failed due to tensile failure, since the load is mainly carried by the plies in the main load direction (0°-direction) and an additional loading of the ±45°-plies was impossible in that case. The predicted failure mode and the load at failure indicate a good agreement with the experimental results. However, the model failed to predict the failure of the 90°-specimens. During the test of the 90°-specimens matrix tensile fracture on the lower side of the specimens was reported, these plies are prone to matrix damage since no fibre are in the direction in the main load. Observing the simulation results of the 90°-specimen indicates that the lower plies failed due to matrix tensile failure, but as elements erosion is only limited to fibre failure the model was not able to predict the collapse of the 90°-specimen.

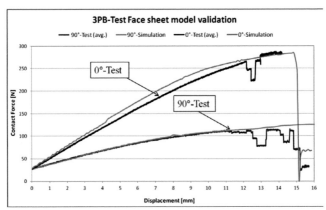

Figure 84. 3PB-Test: Comparison of experimental and numerical results

5.2.5 Foam model validation

To model the crushing behaviour of the closed cell foam the crushable-foam plasticity model with volumetric hardening implemented in Abaqus, which is based on the constitutive model proposed by Deshpande and Fleck [143], was used. To build the yield surface of the foam material the compressive stress-strain curve of the foam and the compression and hydrostatic ratios are required. To obtain the compressive stress-strain curve of the foam ROHACELL® 71 HERO several uniaxial compression tests were carried out. More details of the test execution can be found in section 3.4. The average compression test results used in the simulation are provided in Figure 85. The typical three regions of a compressive stress-strain curve of PMI-foam can be clearly distinguished. At the beginning of the test, the foam behaviour is linear elastic. When the compressive strength is reached, the cell walls begin to crush and buckle at nearly constant load level (collapse plateau). When the cellular structure is completely crushed, the compressive stress increases suddenly because of cell walls and faces coming in contact. This behaviour is known as densification. Combining the Abaqus material model of crushable foam with the linear elastic law would consider these three characteristic regions in the simulation.

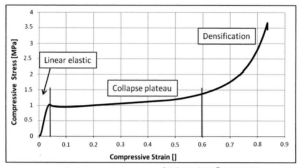

Figure 85. Average compressive stress-strain curve of ROHACELL® 71HERO

In order to determine the compression and hydrostatic stress ratios, foam core indentation tests using a hemispherical steel indenter were carried out. The detailed test setup and results are discussed in section 3.7. Afterwards the quasi-static foam indentation test was simulated with Abaqus. The core was modelled as elasto-plastic material. The initial elastic behaviour was modelled using the elastic isotropic option with the following parameters $E_c = 40.65$ MPa and $v = 0.25$ as inputs. To describe the crushing of the foam the crushable foam model with volumetric hardening available in Abaqus was used, where the stress-strain curve of the uniaxial compression test served as input. The foam specimen (50x50x25.7 mm) was meshed with 8-nodes linear brick elements with reduced integration (C3D8R) and stiffness hourglass control. Since the damage behaviour of the face sheet is mesh dependent the element edge length of the foam model was chosen to be about 0.8 mm, which is the edge length of the elements in the face sheet model, to ensure a similar behaviour during the impact simulation. The indenter was modelled as rigid body with a velocity of 40 mm/s to keep the inertia effects negligible. The same contact condition like in the modelling of the face sheet was used. The compression and hydrostatic ratios were then adjusted until a good agreement between the simulation and experimental results was reached. The indentation model and a comparison between the simulation and the experimental results are provided in Figure 86.

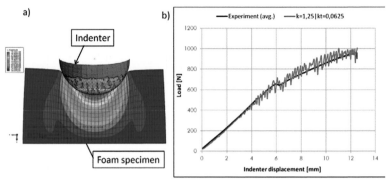

Figure 86. a) Foam indentation model; b) Simulation vs. indentation test results

After validating the two main components of the sandwich structure (the core and the face sheet), the complexity of the simulation model is increased. The component level in the virtual BBA-pyramid is left and the simulation is moved to the assembly level. At this level, the low velocity impact on small and flat sandwich panels is simulated to investigate the interaction between the different sandwich components and the different simulation parameters. At this testing level, the external influence factors on the execution of the impact test with a drop weight impact-testing machine are more controllable compared to testing of large sandwich panels with mobile impactor.

5.2.6 Impact model validation: assembly level

For the validation of the generic impact model flat sandwich specimens were manufactured using the vacuum infusion process described in section 3.1. The test specimens are composed of ROHACELL® 71 HERO foam as core material and CFRP face sheets of about 1.5 mm, made of the materials described in the face sheet model validation section. Two foam core thicknesses were tested: 10 mm and 16.3 mm. The test series is based on plates with the size 350x400 mm clamped around two edges as depicted in Figure 87. All test specimens were impacted using a blunt hemispherical steel impactor with a diameter of 25.4 mm. The impact energy varied between 20 J and 50 J. The impacts were performed at the Fraunhofer IMWS in Halle using the drop-weight impact-testing machine of the type Instron CEAST 9350. For non-destructive evaluation (NDE) air-coupled ultrasound combined with visual inspections were used to determine the failure mode and failure surface. For the validation of the impact model, the contact force versus impactor displacement curves and the failed surface obtained from the simulation were compared to the experimental results. Only the validation of two test specimens with 10 mm and 16.3 mm core thickness respectively that were impacted at 35 J impact energy are presented in this section. More validation results are available in the appendix A3.

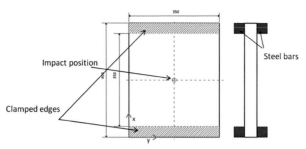

Figure 87. Dimensions of impacted specimen and boundary conditions

A multi-scale modelling approach was chosen to model the impacted sandwich plate with aiming at increasing the accuracy of impact behaviour prediction and reducing the computational time. Hence, the sandwich plate model was divided in three main zones like depicted in Figure 88. In zone one a ply-by-ply modelling approach like in the 3PB-model was chosen to model the upper face sheet under the impactor. The intralaminar failure was considered using the Hashin criterion and the delamination was modelled using the cohesive zone elements like previously explained in the 3PB-test model description. The lower face sheet and the rest of the top face sheet were modelled using one shell element across the thickness without considering a failure criterion as skin failure is unlikely in these zones. The one shell element face sheet was then bonded to the core using tie constraint available in Abaqus package. To simulate the debonding of the face sheet in the top impact area (zone 1) and the bottom interface (zone 1 and 2) cohesive zone elements were used. The cohesive zone parameters were chosen with consideration of the guidelines suggested in [110]. The critical energy release rates G_{IC} and G_{IIC} were assumed equal and were taken from [46]. The strength of the interface is obtained from the strength of the foam core [24], since the crack propagates generally in the foam side of the interface and is controlled by the foam strength [144]. The element edge length varies between 0.8 mm in zone one and 2.3 mm in zone two. A check of the number of cohesive elements in the cohesive zone using the formula proposed by Turon et al. in [109] showed that at least three elements are available in the cohesive zone, which is sufficient for predicting crack growth. Using cohesive elements in the lower interface covering a larger surface than the top interface is meaningful, since cohesive element deletion in the lower interface is a sign of shear cracks in the foam core [7], especially when the foam core model doesn´t consider foam shear crack failure like the used Abaqus material model. Concerning the foam core, the crushable-foam plasticity model with the calibrated parameters from the foam indentation test simulation was applied in the impact simulation. To have an accurate stress distribution under the impactor, a fine mesh was used in the impact zone and zone two with a 1.35 mm edge length in thickness direction. The rest of the model was coarse meshed to reduce computational time. The different zones of the model were then connected to each other using the kinematic coupling method available in Abaqus. The dynamic behaviour of the face sheet and the foam core was assumed rate independent, since rate dependency has only a minor effect on the simulation results of wave controlled impact response as explained in section 5.2.3. The general contact formulation was used to take into account the contact between the impactor, the different plies of the top face sheet and the foam core. The same friction coefficient of 0.3 was assumed for all contact pairs and the impactor was modelled using rigid shell elements. The impact energy was assigned to the impactor by the definition of an impactor weight and an initial impact velocity to the impactor. All degree of freedom of the impactor were constrained except the impact direction to ensure a normal loading of the sandwich specimen.

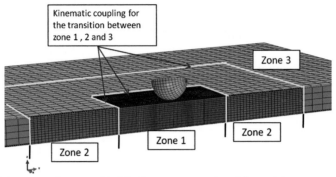

Figure 88. Impact simulation model of the flat specimen: zoning of the model

To evaluate the quality of the simulation model, the peak forces, the failure surfaces and the load displacement curves of the simulation were compared to these obtained from impact tests. Figure 89 and Figure 90 compare the load displacement curves and the specimen failure in the impacted area of two test specimens. The first specimen has a core thickness of 10 mm and the second a 16.3 mm thickness and both specimens showed face sheet rupture with delamination and upper interface debonding as failure modes. Comparing the experimental and numerical specimen cross sections after impact shows a good estimation of the failure occurrence by the numerical model, which predicts the delamination and debonding with high accuracy. The simulation force displacement curves show a good agreement with the experimental results with a slight underestimation of the peak forces and a small over prediction of the impact time and the stiffness of the plate during unloading of the sandwich panel. The failure surfaces of the impacted specimens obtained from the air coupled C-scans correspond to the surface of the debonded interface. The failure surfaces of the simulation were measured from the surface of deleted cohesive elements in the sandwich interface. Table 1 compares the experimental and numerical failure surfaces of the two test specimens taken as examples. The simulation model shows a good agreement with the test results and predicts the failure mode and the failure surface with satisfying accuracy.

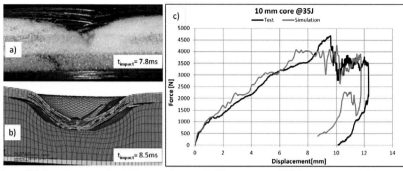

Figure 89. Experiment vs. simulation, specimen with 10 mm core at 35J impact energy: a) Impacted cross section of the test specimen; b) Impacted cross section of the simulation model; c) Force-displacement curves

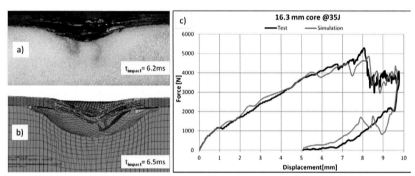

Figure 90. Experiment vs. simulation, specimen with 16.3 mm core at 35J impact energy: a) Impacted cross section of the test specimen; b) Impacted cross section of the simulation model; c) Force-displacement curves

Table 6. Failure surface: Simulation vs. impact test

Test specimen	Simulation [mm²]	Test [mm²]
10 mm core specimen @ 35 J	31x31	37x31
16.3 mm core specimen @ 35 J	29x28	33x28

Using the building block approach to develop a simulation model for predicting the impact behaviour of sandwich panels delivers good results with less iteration loops and reduced effort. Validation of every single component of the sandwich separately and simulating the impact of small and flat sandwich specimens reduces the interaction between the different components of the simulation model and makes the remove of errors and anomalies in the simulation model easier. The next step in the simulation pyramid is to use the gained knowledge and the same approach used to develop the generic impact model to build an impact model for a large aircraft sandwich panel.

5.2.7 Impact model validation: sub-component level

For the validation of the impact simulation model of a large aircraft sandwich panel a 0.8x1.2 m sandwich panel with one flat and one curved face, which represents a section of a vertical tail plane shell of a civil aircraft, was chosen for the impact tests. The radius of the curved face is 6160 mm, which results in maximum core thickness of about 35.5 mm in the centre of the panel and a minimum thickness of about 10.5 mm at the front transition to the monolithic laminate. The face sheets are made with four carbon NCF-layers with a total thickness of about 1.5 mm. The test specimen was manufactured the same way as the generic flat sandwich panels described in the previous section. The sandwich panel was mounted on a support rig made of aluminium. The sandwich panel was fixed at the monolithic run outs along the representative front and rear spar made of CFRP (Figure 91), which simulates the attachment principles and stiffness in a real aircraft tail plane. Full monolithic boosters (Figure 93) with a maximum thickness of about 14.15 mm in the rear booster insure the transition between the run outs and the foam core sandwich.

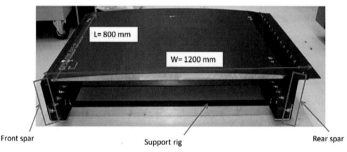

Figure 91. Large sandwich panel and test setup with representative rear and front spar attachments

The impact tests on the large sandwich specimen were performed using a portable impact gun with an impactor mass of 0.735 kg and a diameter of 25.4 mm. The impactor applies impacts normal to the curved surface of the sandwich similar to the drop weight impact system, but uses the stored compressed air as driving force. The impactor displacement, velocity and energy versus time were automatically measured and recorded by the impact gun. The impactor acceleration and the contact force history were calculated afterwards by numerical differentiation. Since the sandwich panel is large enough, interactions between the different impacts were neglected and multiple impacts with impact energies between 25 J and 30 J were performed. The performed impact tests, following the mass criteria [57], can be considered as small-mass or large-mass impacts depending on the affected plate area. Figure 92 shows the performed impacts considered for the model validation. Impact number one was chosen to take into account the interaction with the free edge. Impacts number two and four consider the interaction with the boosters and impact number four doesn't interact with the boundary conditions.

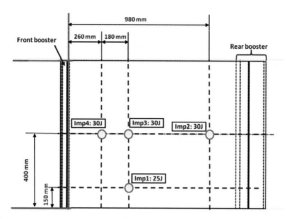

Figure 92. Performed impacts on the sandwich panel

The mesh of the sandwich plate was realised using the Altair pre-processor HyperMesh. The impact simulation and result post-processing were performed with Abaqus. The multi-scale modelling approach was also here used, where the three main zones of the plate like depicted in Figure 88 and the modelling approach for every zone and component like described in the previous section were maintained. The only difference is that the top face sheet in zone two and three and the bottom face sheet in zone three were modelled using the continuum shell elements SC8R available in Abaqus package, in order to reduce the number of interaction constraints in the model. Because of the complex geometries of the boosters, the use of continuum shell elements with composite properties was not possible. 3D-solid elements (C3D8R) with orthotropic material properties were employed instead. The test frame and front and rear spars were not considered in the simulation model. It was assumed that the joint between the panel and the test frame is rigid and the impact behaviour is not affected by the vibration of the support frame, hence the monolithic run outs in contact with the front and rear spar were fully constrained at the nodes highlighted in Figure 93. The impactor was modelled with rigid shell elements and the Abaqus general contact definition was used. Pre-damages by previous impacts were not considered in the simulation and it was assumed that they have only minor influence on the subsequent impact, so that they could be neglected in the simulation.

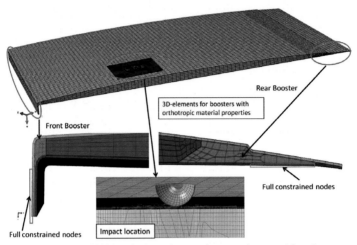

Figure 93. Impact model highlighting the boundary conditions and rear and front boosters

The measured and simulated load and displacement histories are depicted in Figure 94. In spite of the noise in the experimental data, the predicted and measured impact data show generally a fair agreement. The best agreement between experiment and simulation is observed in impact number one, which shows a good prediction of the impact response during all impact phases. The other impacts overestimated the pick load and the impact time slightly. The stiffness of the plate during the loading is well predicted, while the stiffness is underestimated during the unloading phase. Regarding the load histories from the simulation the first change in the load history occurred at about 800 N, at this load level, the stress in the core exceeded the compression strength of the foam and the foam crushing initiated. The second characteristic point is at about 1600 N, the simulation showed that the delamination began at this load level. The next change in plate stiffness is at about 3400 N, at which interface debonding and first ply rupture were observed. At the pick load, multiple plies failed at the same time, which resulted in clear load drop and stiffness degradation. These characteristic points were also observed during the simulation of the generic flat sandwich specimens discussed in the previous section. Nearly the same loads for core crushing, delamination and face sheet rapture were also found in the study of the indentation behaviour of the same sandwich configuration reported in section 3.6, which confirms the similarities between the indentation and impact behaviour of foam core sandwich.

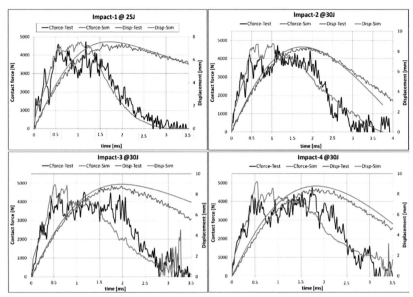

Figure 94. Comparison of measured and predicted load and displacement histories

Considering the velocity histories depicted in Figure 95 the simulated and measured data appear to be in good agreement with the exception that, the reversal point, at which the impactor velocity is zero and the impactor changes its direction, was reached slightly earlier during the simulation compared to the test results.

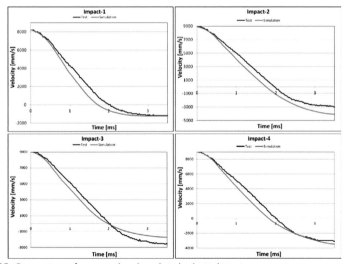

Figure 95. Comparison of measured and predicted velocity histories

111

During the impact tests only face sheet rupture with foam crushing were observed. The simulation predicted the same failure modes with underestimation of the failure surface. The deviation from the experimental results is larger compared to the deviation observed in the impact simulation of the generic flat panels. This deviation could lie on the assumption that the test fixture is rigid, while it was not the case during the test, and the precision of the portable impact gun is lower than the precision of the drop-weight impact-testing machine.

Table 7. Failure surface: Simulation versus impact test of the large panel

Impact number	Simulation [mm²]	Test [mm²]
Imp-1	22x20	16x16.6
Imp-2	27x25	17.5x17
Imp-3	26x27	23.5x21.2
Imp-4	21x20	17.4x18.7

5.3 Numerical simulation of impact on TFC-sandwich structure

In this section, a finite element model for the prediction of the impact behaviour of pin-reinforced sandwich structures is presented. The same approach based on the building block approach described in section 5.2 was adapted to develop the model. Multi-scale modelling in the impact region that considers the delamination of the face sheet using cohesive zone elements was employed to increase the accuracy of the simulation. The impact model was validated by comparing the FEM-results with impact test results. Moreover, two different approaches to model the pin-reinforced foam core were investigated, analysed and compared. The first approach is a homogenisation approach that considers the pin-reinforced foam core as a homogeneous entity, which means the pin-reinforced core is modelled with 8-node brick elements with the properties of the TFC-core obtained from the mechanical tests. In this approach the pins are not modelled. The second approach is a detailed approach that considers the pins in the core model, as they are modelled with 8-node brick elements and the surrounding foam is modelled using 4-node brick elements with standard foam material model.

5.3.1 Literature review

To enable the integration of foam core sandwich into primary aircraft structure, the impact tolerance should be assured and the impact behaviour at defined conditions should be predictable. The finite element method presents a reliable mean to predict the impact behaviour of sandwich structures. While most of the researchers used the experimental methods to investigate the impact response of pin-reinforced foam core sandwich structures, only few studies focused on the numerical investigation of the impact behaviour of stitched foam core sandwich. Xia et al. [145] developed a model to predict the impact damage of through-thickness stitched foam core sandwich structure. The model was built with consideration of symmetry conditions of the sandwich panel and using 2D-solid element. The Kevlar fibre columns vertically inserted into the foam core were considered in the model. While acceptable damage prediction was reached, the accuracy to predict the force-displacement curves was poor. Han et al. [85] investigated the effect of impact energy and stitch density on low-velocity impact behaviour of foam core sandwich panels with glass fibre stitches. A 3D-model was developed that simplified the stitching region as glass fibre reinforced resin columns. Connecting both face sheets with the stitches was neglected. A strain based continuum damage mechanics approach was used to predict the damage occurrence in the face sheets. The crushable foam core model was used for the foam material. A good agreement between simulation and experimental results was generally reached for the investigated impact energy range. But increasing the impact energy would probably increase the damage in the core and would bring the impact model to its limit.

Since the pins are not always inserted vertically into the foam core and complex unit cell geometries are usually used, modelling the pin reinforced foam core will be more complicated and this would increase the modelling effort especially for large sandwich structures. In addition, increasing the impact energy leads to more damage in the core, which could lead to simulation troubles like excessive mesh distortion and high stiffness mismatch between the foam core and the rigid pin reinforcements. Alternative modelling approaches like considering the pin reinforced foam core as a homogeneous entity or developing new material models could resolve this problem.

5.3.2 Modelling approach

To develop a reliable prediction of the impact behaviour of pin-reinforced sandwich panels the same approach based on the BBA and described in section 5.2.2 was applied. The model of the face sheet was previously validated in section 5.2.4 and the same parameters are used in this section. The different pin-reinforced core were validated by simulation of quasi-static tests (flatwise compression and indentation test) and comparing the simulation results with the experimental data. Two different modelling approaches were compared and a suited approach was chosen for the impact model. Afterwards the impact model was built using the parameters from the validated coupon models. The simulation results were compared to impact test results in terms of impact force-displacement diagrams and failure modes in order to validate the simulation model. Following this approach enables to validate the impact model with reduced effort and to expedite the elimination of anomalies in the simulation.

5.3.3 Validation of the TFC-model

To model the pin-reinforced foam core two different approaches have been chosen. The first approach, which is called the homogenisation approach, considers the pin reinforced foam core as a homogeneous entity consisting of one material with the properties obtained from the flatwise compression and indentation tests. Only 8-node brick elements are needed to model the TFC-core without need to model the pins, so that the modelling effort is extremely reduced. The second approach considers every details of the unit cell geometry except the pin extensions. The pins are modelled using 8-node brick elements having the properties of the pins and the surrounding foam material is modelled with 4-node brick element assigned to the foam properties. The second approach requires intensive efforts to model the pin reinforced core, as the complexity of the geometry increases with the increase of the pin density.

5.3.3.1 The homogenisation approach

The homogenisation approach is based on the black box concept, the pin reinforced core is considered as homogeneous material with the properties obtained from the flatwise compression and indentation tests. The internal interaction between the foam material and the pins is not considered. As the overall behaviour of the TFC-core doesn´t really differ from the behaviour of the unreinforced foam material, following the analysis in section 3.4, the linear elastic law for isotropic materials was used to describe the initial linear elastic behaviour of the core, the Abaqus material model "crushable foam core with volumetric hardening" was chosen to model the crushing of the pin reinforced foam core. The compressive modulus, compression yield stress and the uniaxial plastic strain measured during the quasi-static tests were used as input parameters. The compression yield stress ratio and the hydrostatic yield stress ratio could be obtained iteratively by simulating the foam core indentation test. The flatwise compression test was simulated to validate the uniaxial compressive behaviour of the core. The model consisted of a discrete rigid flat panel loading the top face of the test specimen in compression. The penalty contact method with a friction coefficient of 0.3 was chosen to model the contact between the loading plate and the test specimen. The lower face of the specimen was constrained in thickness direction. As starting values for the compression and hydrostatic yield stress ratios the following ratios were used: $k=1.5$ and $k_t=0.075$. The second calibration test was the foam core indentation, the model consisted of the same features of the uniaxial compression test model, with the exception that a rigid hemispheric indenter ($\phi=25.4$ mm) was modelled instead of the rigid plate. Both models are depicted in Figure 96.

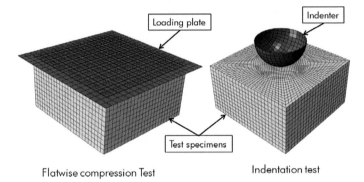

Flatwise compression Test   Indentation test

Figure 96. Homogenisation approach: used models for TFC-validation

Figure 97 shows the simulation results compared to the experimentally determined load curves of the flatwise compression test. The applied material model was able to predict the three different zones of the stress-strain curves with high accuracy. The compressive strength of the tested TFC-configurations and the rising plateau stress were precisely determined by the simulation model. In spite of neglecting the interaction between the foam and the pins the material model "crushable foam core with volumetric hardening" seems to be appropriate to describe the uniaxial behaviour of pin reinforced foam core correctly.

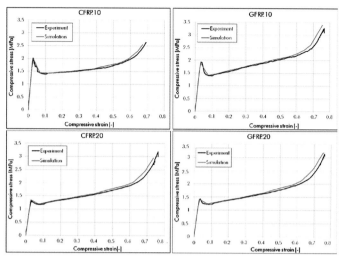

Figure 97. Homogenisation approach: flatwise compression test, simulation vs. experiment

The second calibration test is the foam core indentation, which enables to investigate the interaction of compressive and shear loads when the steel indenter is pushed into the core. The contact interaction between the pins and the indenter was not considered as the pin reinforced core was assumed to be homogeneous. For the calibration of the core model the input data from the flatwise compression test, namely the compression yield stress and the uniaxial plastic strain, were not changed. Several iteration steps were performed, in which the compression yield stress and hydrostatic yield stress ratios were iteratively changed, until an agreement between the experimental and simulation load curves was reached.

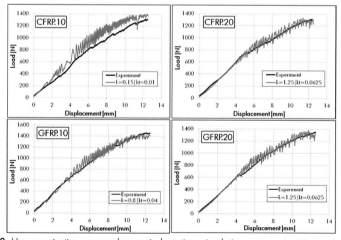

Figure 98. Homogenisation approach: core indentation, simulation vs. experiment

115

An overall good agreement between the simulation and the experiment can be shown in the load-displacement curves depicted in Figure 98. The simulation shows generally an accurate correlation in the initial linear region and in the load softening region when the foam cells crushing started. The exception is configuration CFRP.10, which shows a stiffer indentation response in the simulation. Despite further iteration steps no additional improvement of the simulation results of configuration CFRP.10 was reached. The determined compression yield stress and hydrostatic yield stress ratios are listed in Table 8.

Table 8. Determined compression yield stress and hydrostatic yield stress ratios

Configuration	k [-]	k_t [-]
CFRP.10	0.15	0.01
CFRP.20	1.25	0.0625
GFRP.10	0.80	0.04
GFRP.20	1.25	0.0625

During the indentation test the energy is mainly consumed by foam core crushing. At a defined indenter displacement shear cracks are created under the indenter (Figure 52). When pins are present in the core pin fragmentation occurs below the indenter as well. These kinds of damages cannot be simulated with the used material model. In spite of neglecting the internal interactions in the core, the material model seems to be suited to simulate the global behaviour of the pin reinforced foam core under indentation. This kind of loading can be also observed below the impactor when a sandwich structure is subjected to low-velocity impact. The suitability of the model to describe the impact behaviour of pin reinforced foam core sandwich will be further investigated in section 5.3.4.

5.3.3.2 The detailed approach

This modelling approach considers as much as possible details to simulate the mechanical behaviour of the pin reinforced foam core. The main advantage over the homogenisation approach is that only the mechanical properties of the foam and pin materials are needed to estimate the mechanical properties of the pin-reinforced foam core. A successful modelling of the mechanical behaviour with the detailed approach should reduce the testing effort. In this model the pins were modelled using 8-node brick elements with orthotropic elastic properties. The fibre volume contents of the carbon and glass fibre pins were determined previously and the properties were calculated using rule of mixtures and manufacturers properties. The used material properties of the pins are listed in Table 3. Due to the restrictions of the used material model and the complex geometry of the unit cell, only the elastic behaviour of the pins was considered and the damage initiation and propagation in the pins were omitted. The pin extensions were neglected due to high modelling cost. The pins were connected to the surrounding foam using a node-to-node connection assuming a perfect connection without failure. The foam ROHACELL® 71 HERO was modelled using the Abaqus material model "crushable foam core with volumetric hardening" and the validated material parameters from section 5.2.5. Due to the complex geometry, 4-node brick elements were used to model the foam material. Figure 99 shows an exemplary representation of a pin reinforced foam core with 10 mm pin distance and modelled using the detailed approach. To verify the validity of this approach the flatwise compression and the indentation tests were simulated and the numerical results were compared to the experimental data. Unlike the homogenisation approach no iteration steps to determine the compression yield stress and hydrostatic yield stress ratios are needed, as the effect of the pins on the mechanical properties is considered when the detailed approach is used. Hence for every pin configuration only a single simulation run is needed to verify the indentation behaviour.

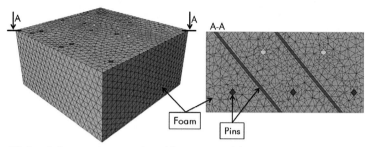

Figure 99. Detailed approach: pin reinforced foam core model

Figure 100 compares the simulation with the experimental results of flatwise compression test. It can be seen that the model is able to predict the initial linear elastic stiffness of all configurations with high accuracy. But, it failed to estimate the compressive strength and the plateau stress of the tested configurations. This imprecision lies on the neglect of pin damage and the interaction between the pins and the foam after damage initiation like it is shown in Figure 34. Moreover, the experimental analysis of the damage modes and analytical model in section 3.4.4 have shown that the pin compressive strength is the most important factor that influence the compressive strength of pin reinforced foam core sandwich structure.

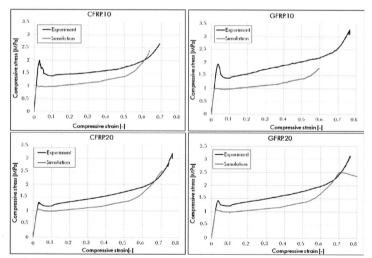

Figure 100. Detailed approach: flatwise compression, simulation vs. experiment

From the comparison presented above it can already be seen that the proposed detailed approach is not suited to simulate the mechanical behaviour of the TFC-sandwich. The pins damage behaviour and the interaction with the foam should be considered to improve the simulation results. For the sake of completeness the comparison of the foam core indentation test results to the simulation results is presented in Figure 101. The models with 10 mm pin distance (CFRP.10 and GFRP.10) failed to finish the simulation run due to excessive elements distortion. Remeshing of the model with smaller elements or other meshing methods didn't deliver any improvement. Nonetheless, a good accuracy in the linear elastic region can be observed in the load-displacement diagrams of CFRP.10 and GFRP.10. Concerning the configurations CFRP.20 and GFRP.20 with 20 mm pin distance, the simulation was completed until the end of the simulation step without any interruption. However, the simulation results show a reduced stiffness in the initial linear elastic section of the load-displacement diagram and a softer behaviour at crushing compared to the experiments. These observations confirm the conclusions retained from the simulation of the uniaxial compression, that this approach with the prior made assumptions is not suited to model the impact behaviour of TFC-sandwich.

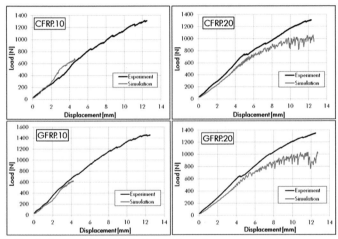

Figure 101. Detailed approach: core indentation, simulation vs. experiment

In addition to the unsuccessful prediction of the load-displacement curves under uniaxial compression and indentations loads, the elements of the detailed approach model showed excessive distortion compared to the model built with the homogenisation approach (Figure 102). This lies on the stiffness jump and the perfect bonding (node-to-node connection) between the foam material and the pins. This excessive deformation can lead to the interruption of the simulation due to elements distortion, if the sandwich panel is subjected to high local deformations as result of low-velocity impact.

In order to get useful simulation results with the detailed approach it is necessary to consider the damage evolution in the pins and the collapse of the bonding interface between the pins and the foam material. However, considering all these damage types would increase the simulation effort and would make the use of this approach to simulate the impact behaviour of TFC-sandwich unreasonable.

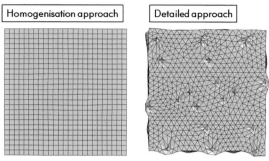

| Homogenisation approach | Detailed approach |

Figure 102. Comparison of elements deformation

5.3.4 Validation of the impact model

Following the results of the TFC-model validation an impact model using the validated face sheet properties from section 5.2.4 and the homogenisation approach for the pin reinforced foam core was developed. The model should be able to predict the damage mode, the damage surface and the contact force-displacement diagram with good accuracy. Using the homogenisation approach to model the TFC-core enables to use the same model of the foam sandwich panels described in section 5.2.6. Only the geometry of the panel and the material properties of the core should be modified, the remaining parameters should remain unchanged. The multi-scale modelling technique was used as well to model the sandwich panel, which was sectioned in three main zones: zone 1 is finely meshed with ply-by-ply modelling of the top face sheet, zone 2 and 3 are coarser meshed without damage simulation to ensure the panel stiffness with reduced computation costs. The different zones were then connected using the kinematic coupling method available in Abaqus/explicit. The debonding of the face sheet was also considered by the use of cohesive zone elements connecting the face sheets to the foam core in zone 1 and 2. The critical energy release rate in normal and shear loading directions were assumed to be equal and were taken from the work of john et al. [46], who performed single cantilever beam test to determine the critical energy release rate G_{IC} of the interface of sandwich with CFRP.10 and CFRP.20 pin pattern. The impact energy was applied by assigning mass and initial velocity to the rigid impactor. The impact setup and a cross-sectional view of the simulation model are provided in Figure 103.

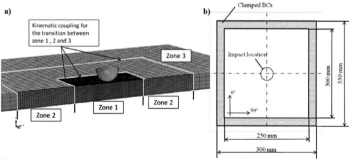

Figure 103. a) Impact simulation model; b) Specimen and impact boundary conditions

For the purpose of model validation six specimens with the size 350x300 mm were manu-factured and impacted using the same testing device described in section 4.2. Only specimens with pin configurations CFRP.10 and CFRP.20 were tested for validation pur-pose. The sandwich panels were fixed within a picture frame with an impact window of 250x300 mm and impacted in the centre with a hemispherical steel impactor of diameter 25.4 mm at room temperature. Afterwards ultrasound C-scan was performed and some specimens were sectioned to determine the failure mode and the damage surface. The core thickness varied from 10 mm to 16.3 mm. The face sheets had a thickness of 1.5 mm or 2.25 mm. All of these specimens were subject to face sheet rupture except specimen P2.1 with 10 mm core and 1.5 mm face sheet thickness, which was subject to foam core shear failure in addition to the face sheet rupture. The geometry of the panels, the impact energy and the resulted damages are listed in Table 9. The test results were compared to the numerical results in terms of impact force-displacement diagrams, failure modes and damage surfaces.

Table 9. Summary of the test series and resulted damages

Panel	Core thickness [mm]	Face thickness [mm]	TFC-type	Energy [J]	Damage
P1.1	10.00	1.50	CFRP.20	35	face sheet
P1.2	16.30	2.25	CFRP.20	35	face sheet
P1.3	16.30	2.25	CFRP.20	50	face sheet
P2.1	10.00	1.50	CFRP.10	35	core and face sheet
P2.2	16.30	2.25	CFRP.10	35	face sheet
P2.3	16.30	2.25	CFRP.10	50	face sheet

The force-displacement diagrams of the simulated impacts, displayed in Figure 104, show an overall good agreement of numerical results and tests. The initial linear stiffness, the beginning of face sheet degradation as well as the dissipated energy and the unloading phase of the impact were estimated with high accuracy. However, the peak force for face sheet rupture and the following plateau load were not always correctly guessed, but the deviation from test results lay always in acceptable tolerance. This could lay on the fact, that face sheet waviness is created by the pin extensions, which creates punctual thickening of the face sheet and as a result the face sheet rupture load depends strongly on the im-pact location and can deviate from the theoretical value predicted with the Hashin crite-rion. In addition, the pin reinforced foam core experiences different damage modes than the neat foam, which cannot be considered by the used foam material model; as a result deviations could be observed in the estimated plateau load. Generally, impact tests don´t always deliver repeatable results, some deviation can be observed as the impact response is very sensitive to the manufacturing quality and the impact boundary conditions. As the simulation effort should be kept at a reasonable level, it was not possible to consider all these small details, but nevertheless the impact response estimated by the proposed model correlates reasonably well with the experimental results.

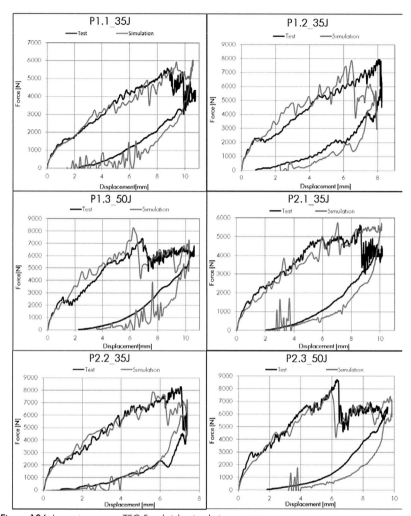

Figure 104. Impact response TFC-Sandwich: simulation vs. test

Considering the damage mode, the proposed model was able to predict the damage modes of the six impact panels correctly. A first glance at the impacted area from cross-section view of panel P2.2 depicted in Figure 105 shows many similarities of the damage pattern of the face sheet and the core material beneath the impactor. The Hashin criterion combined with the cohesive element law exhibited similar face sheet damage pattern.

Generally, when a shear crack initiates in a foam core it propagates until reaching the lower face sheet and then continues to propagate in the lower interface creating face sheet debonding. Since the used foam core material model doesn´t consider shear cracks, this damage mode was identified by the deletion of the cohesive elements in the lower panel interface (Figure 106), like suggested in [7]. However, as the used approach to model pin-reinforced core doesn't consider the shear damage mode and no further verification with panels with core shear damage was performed, it is recommended in addition to the simulation to use analytical methods, like the model proposed by Olsson and Block [56], to verify the initiation of core shear cracks.

Figure 105. Cross section of panel P2.2 after impact: test vs. simulation

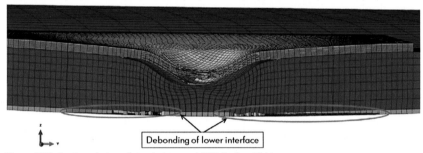

Debonding of lower interface

Figure 106. Sectioned view of simulation model P2.1 after 35J impact

Table 10 compares the experimentally determined impact duration and damage surface to the values estimated by the simulation. Two accuracy levels can be distinguished: the simulation of the panels with CFRP.20 pin pattern shows good accuracy, only small deviations can be observed. The deviation from the experimental results increased with the simulation of the panels with CFRP.10 pin pattern, especially the damage surface of the panel P2.1 with core shear damage deviates strongly from the experimental determined value. This failed estimation shows the limitation of the used material model for the pin reinforced foam core due to no consideration of the shear damage mode by the core material model. These two level accuracy of the simulation results could be explained by the influence of the pins on the mechanical behaviour. The influence of the pins on the mechanical behaviour of the panels with CFRP.20 pin pattern is limited compared to the panels with 10 mm pin distance (CFRP.10). The experimental investigations in chapter 3 showed that the mechanical properties of the core with pin configuration CFRP.20 don't differ strongly from the properties of the unreinforced foam core, which explains the good accuracy of the impact simulation.

Even if the shear damage surface was strongly underestimated for panel P2.1, the model is still useful to predict the core shear damage mode, which is important to avoid critical sandwich designs. Further improvement of the impact simulation could be obtained by the implementation of the shear damage in the core material model and calibration of the cohesive elements parameters of the sandwich interface.

An exact simulation of the low-velocity impact is generally impossible, but a minimum of accuracy is needed to deliver reliable estimations of the impact response, which is provided by the proposed impact model.

Table 10. Comparison of the estimated impact duration and damage surface to the experimental values

Panel	Impact duration [ms]		Damage surface [mm^2]	
	Test	Simulation	Test	Simulation
P1.1	8.1	7.8	42x36	46x33
P1.2	6.6	6.5	44x36	40x28.5
P1.3	8.8	8.8	40x38	41x33
P2.1	8.0	7.6	158x156	68x41
P2.2	6.5	5.8	64x44	33x34
P2.3	8.5	7.6	46x43	37x34

5.4 Chapter summary

The present chapter consists of two main sections: in the first section the low-velocity impact behaviour of a sandwich structure with complex geometry representative for a large-scale aircraft sandwich panel was simulated and validated using a BBA-based simulation strategy. In the second chapter section a simulation model for pin reinforced foam core was developed and validated, whereby two modelling approaches were studied and compared, namely a homogenisation and a detailed approach. Then the chosen pin reinforced foam core model was applied to the impact model of pin reinforced foam core sandwich panel developed in the previous section.

This two-level modelling strategy of the impact behaviour of the TFC-sandwich was chosen with the aim to gain know-how about low-velocity impact modelling of foam core sandwich panels and to adjust the simulation parameters, as well as to reduce the effort of incorporating the pin-reinforced foam core model into the impact model.

In the first section a strategy based on the building block approach, which is typically used to design composite structures for aerospace applications, was chosen to develop the simulation model for impact on aerospace sandwich structure. The chosen panel for model validation is representative of a vertical tail plane side shell of a civil aircraft. The sandwich panel considers the aerodynamic curvature of the top face sheet, the attachment to the front and rear spars and the transition between the monolithic run outs and the sandwich section. First, the models of the face sheet and the foam core were validated separately by comparing the numerical results to experimental results obtained from quasi-static tests. In second step, an impact model for generic sandwich panels was developed using the material properties validated in the previous step. Impact test on small flat sandwich panels were performed using a drop weight tower. The load-displacement curves and the failure surface obtained from US-scans were used to validate the generic impact model. Finally, the material and simulation parameters used to simulate impact on small flat panels were used to model the aircraft sandwich panel. Impact tests using a portable impact gun were performed to generate impact data for model validation.

A good agreement between the experimental and numerical results was found, the model predicts the failure occurrence and the load-displacement curves with high accuracy. The applied modelling strategy yields to fast validation of the impact model of large sandwich panel without extensive model adjustment or uncertainty when adjusting the material and model parameters. The only drawback of this modelling strategy is that enough experimental results should be available at different simulation levels to validate the simulation results. The used modelling strategy can be applied to other complex geometries of sandwich structures and would lead to satisfying results.

In the second section two modelling approaches for the pin reinforced foam core were studied and compared. The first approach, "the homogenisation approach", considered the TFC as a black box with homogeneous properties. The interaction between the foam and pins was not considered and the Abaqus material model "crushable foam core" was used. The flatwise compression test and the core indentation test were simulated and the material model parameters were adjusted until a good agreement was reached. The second approach, "the detailed approach", considered the pins and assumed a perfect connection with the foam material. The Abaqus material model "crushable foam core" was applied to the foam material with the validated parameters from the first section of this chapter, and orthotropic material properties calculated with the rule of mixtures were applied to the pins. This model is marked by extensive modelling effort and long simulation time compared to the homogenisation approach. Moreover, the model failed to predict the core behaviour under flatwise compression and indentation loads and some simulations were interrupted due to excessive elements distortion. For these reasons this modelling approach was not integrated in the impact model and only the model built with the homogenisation approach was considered.

An overall good agreement was reached with the proposed impact model of the TFC-sandwich. The damage was always predicted correctly and the estimation of the force-displacement curves showed reasonable accuracy. Considering the estimation of the damage surfaces and the impact duration, two level of precision could be observed. Only minor deviations were observed for the panels with 20 mm pin distance (CFRP.20), while the deviations increased for the panels with 10 mm pin distance (CFRP.10) especially when core shear cracks occurred after impact loading. The increase of the deviation can be explained by the fact that the used material model doesn´t consider the core shear damage mode and the increase of the pin volume fraction in the core increases the influence of the pins on the mechanical behaviour of the core, which reduces the suitability of the used material model to this kind of foams with high pin volume fraction.

The low-velocity impact behaviour of foam core sandwich structures is complex and depends strongly on the properties of the used materials, the boundary conditions and the sandwich geometry. Adding pins to the foam core manipulates the mechanical properties and increases the challenges when impact resistant sandwich structures are designed. The finite element method is suited to design such kind of sandwich structures, but experiment based validation strategy should be applied and the limitations of the assumptions and the model should be known to avoid the misinterpretation of the simulation results.

6 Summary and outlook

Damage tolerance is a major criterion to design primary aircraft structures. Invisible damages in the structure should not reduce the required performance and endanger the integrity of the structure. As sandwich structures are vulnerable to out-of-plane loading and invisible damages are likely to occur it is important to study the damage onset and propagation under different loading scenarios. Adding pin-reinforcements to foam core sandwich structures would improve the impact resistance and damage tolerance and enhances the mechanical properties of the foam, leading to reach performance level near the honeycomb mechanical performance. But, the core reinforcement creates new materials interactions and increases the complexity of the mechanical analysis and structure design process. In order to understand the behaviour of pin-reinforced foam core sandwich structures and to cover all types of damage onset and propagation, extensive experimental investigations are required and new analytical and numerical methods are needed. In this work, the mechanical response of pin-reinforced foam core sandwich structures under quasi-static compression, shear and indention loading conditions was investigated with focus on damage onset and propagation. In Addition, the impact response of the same structures at very severe frost conditions (-55 °C) was studied and new damage modes and factors of influence were found out. Moreover, simulation models to predict the impact behaviour of complex shaped foam core sandwich structures and pin-reinforced foam core sandwich panels were developed and validated against experimental results.

In chapter 2, an overview about the applications of sandwich structures and fundamentals of composite sandwich structures is given. The topic of pin-reinforcement of foam cores was discussed in detail, where the motivation behind the principle of pinning and the different pinning technologies were presented and the tied foam core technology was introduced. The last part of the chapter deals with the topic of impact behaviour of foam core sandwich structures with focus on the typical impact damage modes and damage onset criteria. It was shown that is necessary to use analytical methods in early design stage in order to reduce development time and to avoid critical shear cracking. The sandwich configuration should have a critical load for core shear cracking higher than the face sheet rupture load in order to avoid core shear damage. A minimum core thickness for a given face sheet is required to increase the critical shear load and to minimize the risk of core shear cracking.

In chapter 3, the TFC-sandwich structure was characterised by means of flatwise compression, out-of-plane shear and indentation tests. Specimens with glass fibre and carbon fibre pins and two different pin-volume fractions were tested. The micro computed tomography was used to identify the pins damage behaviour. The following conclusions can be identified:

- The pins integration led to the improvement of the compression, shear and indentation behaviour.
- An obvious improvement was firstly detected at 0.72% pin volume fraction. The effect of the lower pin-volume fraction (0.2%) is minor.
- The glass fibre pins led to a higher crushing energy during flatwise compression tests than the carbon fibre pins.
- Bending damage of the pins at the curved pin ends is the first pin damage mode that occurred during flatwise compression and indentation tests.

- The specimens with 0.72% pin volume fraction showed better structural integrity after core shear cracking and the pins led to crack deviation and stopping, while the behaviour of the specimens with 0.2% pin volume fraction was similar to the behaviour of the unreinforced foam core.

Later in chapter 4, the low-velocity impact behaviour of pin-reinforced foam core sandwich structures was investigated. It was found that using carbon fibre pins with a pin volume fraction of about 0.72% can lead to the creation of thermal induced cracks in the core when impacted with 35 J impact energy. These cracks have an undefined propagation pattern and were not observed during impact tests at room temperature. The creation of thermal cracks may lay on the difference between the thermal expansion coefficients of the foam core and the carbon fibre. Two methods were investigated to delay the thermal cracks to higher impact energies. Sandwich panels with carbon fibre pins were manufactured with vacuum assisted resin infusion process in the autoclave using the autoclave pressure to reduce the thermal expansion of the core and the thermal stresses. The second methods consisted of using glass fibre pins instead of carbon fibre pins. Both methods led to the delay of the onset of core thermal cracks to higher impact energies. The problem of thermal induced cracks in the foam core was not observed in the unreinforced sandwich panels. This means that the pins added a new constraint that should be considered during the design process.

The impact simulation model was developed in chapter 5. In the first section of the chapter, the modelling approach based on the building block approach for aerospace structures was validated for unreinforced sandwich structure. The modelling and simulation started from element level with the validation of the face sheet and the foam core models until reaching the detail/demonstrator lever in which the impact behaviour of a complex foam core sandwich panel was simulated. The first part of the simulation served to verify the modelling approach, to validate the simulation parameters, as no pins in the core means less model calibration, and to demonstrate the feasibility to simulate the low-velocity impact behaviour of large sandwich structures with complex geometry by using multi-scale modelling approach. In every simulation step the numerical results were compared to experimental results and validated. A good agreement was achieved in every step.

In the second part of chapter 5 the low-velocity impact behaviour of flat pin-reinforced foam core sandwich panels was simulated. Two approaches to model the pin-reinforced foam core were investigated. In the first approach the pin-reinforced foam was considered as homogenous entity so that the pins were not modelled and the test data from the flatwise compression and indentation tests served as input parameters to the material model. The model doesn't consider the interaction between the foam core and the pins, and the material parameters should be calibrated by the simulation of quasi-static test. In the second approach the pins and the foam core were modelled assuming an ideal bonding between them without pin debonding. Due to simulation limitations, the pin extensions and the damage behaviour of the pins were neglected, and the mechanical properties were calculated using the rules of mixture. Moreover, this approach increases the modelling effort as every single pin should be modelled. While a good agreement with the experimental results was reached with the homogenous model, no model validation with the second modelling approach was possible as the core strength was underestimated and the simulation aborted several times due to excessive elements distortion. The model with homogenous core properties was extended to simulate the impact behaviour of flat sandwich panels with pin-reinforced foam core using the same multi-scale modelling approach of the previous section. The model succeeded to predict the damage modes of the simulated impact cases. A good accuracy was reached for moderate impact damages, but the damage surface was underestimated for shear crack damage and large sandwich damage. This deviation may lay mainly on the core cracks that are not considered in the core model and the extensive pin damage at high impact energy, which affects the debonding surface. The proposed model gives the possibilities to study different impact scenarios and provides good accuracy especially for impact cases with moderate damages. The model can build a virtual test environment and enables to reduce physical tests as critical test cases can be virtually detected. In order to improve the result accuracy and to extend the validity of the model, the following improvement are necessary:

- Material model for face sheets made of non-crimp fabric layers is needed.
- Properties of the face sheet material should be obtained from tests on NCF-coupon tests not from assumptions based on similar UD-layers.
- A model for the foam core that simulates core cracking and damage evolution.
- Consideration of the pins in the foam core and using a material model that simulates the different pin damage types.
- Validation of the cohesive elements in the interface zone between the core and the face sheet based on debonding tests.
- Consideration of residual stresses in the core induced by the manufacturing process.

In this work, it was shown that adding pin reinforcement into the core improves the mechanical properties of the sandwich structure, but it creates new damage types and new interactions between the different sandwich components that should be considered during the design process. For integration in primary aircraft structure the designed sandwich panel should be damage tolerant, which means a special focus on the invisible damages like core cracks should be put. The performed low-velocity impact tests at -55 °C revealed that thermal cracks are likely to occur in the foam core if unsuitable materials combination is used. To avoid unwanted surprises it is important to verify the impact behaviour of pin-reinforced foam core sandwich structures at the lowest service temperature in early design stages, especially if the structure is used in very cold conditions. An analytical model that predicts the thermal residual stresses in the pin-reinforced foam core would enable to make a better decision when choosing the materials and would reduce the development time. In consequence of the impact results, it is recommended to perform cyclic mechanical and thermal fatigue tests to verify the long term structural response of the pin-reinforced foam core sandwich structure, as the pins may create local stress concentrations due to notch effect in the foam material. Unlucky choice of the foam core type can lead to the reduction of the structure life time. Moreover, to gather more detailed knowledge of the damage tolerance of the structure performing compression-after-impact tests would deliver valuable findings.

In this work the considered impact scenarios for the pin-reinforced sandwich correspond to the large-mass impact response type. The investigation of impact response of small mass impact on large sandwich panels would be very beneficial to have a complete understanding of the behaviour of the pin-reinforced foam core sandwich structure as the small-mass impact type represents other relevant impact scenarios like as e.g. runway and engine debris or hail strike, which have high impact velocities that lead to important strain rate effects.

The foam core pin-reinforcement is based on the idea to improve the impact resistance and the damage tolerance by reducing the amount of damage through crack deviation and stopping, creating new load paths and increasing the structural integrity by connecting the face sheets. This method showed that it is not possible to avoid the core damage completely and led to the creation of new damage types at very low testing temperature (-55°C). Another concept to improve the impact resistance of foam core sandwich structure was tested in another study [146]. The concept is based on the idea to transform the impact energy into rebound energy rather than into damage dissipation energy. By adding a thin material layer with rubber-like behaviour between the foam core and the face sheet the rebound level of the impactor would increase and the impact damage would reduce. This concept was verified by performing impact tests on reference foam core sandwich panels and panels with a rubber-like layer. Sandwich panels were manufactured with VARI-process and a thin cork layer was integrated between the foam core and the face sheet on the impact loaded side of the panel. The performed impact tests showed that it is possible to suppress core shear damage for the investigated impact energy range and higher amount of the impact energy was transformed into impactor rebound energy compared to the reference panels without cork.

References

1. Zuardy MI, Block TB and Herrmann AS. Structural analysis of center box of a vertical tail plane with a side panel from composite foam sandwich. In: Proceedings of Deutscher Luft- und Raumfahrtkongress, Bremen, Germany, 2011; pp. 1377–1389.

2. Herrmann AS, Zahlen P and Zuardy MI. Sandwich structures technology in commercial aviation. In: Proceedings of the 7th International conference on sandwich structures, Aalborg, Denmark, 2005; pp. 13–26.

3. Heimb S. Energy absorption in aircraft structure. In: International workshop on hydraulic equipment and support systems for mining, Huludao, China, 2012.

4. Stanley LE and Adams DO. Development and evaluation of stitched sandwich panels. NASA technical report no. NASA/CR-2001-211025, 2001.

5. Frederick G, Kaepp G, Kudelko G et al. Optimization of expanded polypropylene foam coring to improve bumper foam core energy absorbing capability. SAI technical paper no. 950549, 1995.

6. Fasanella EL, Jackson KE, Lyle KH et al. Multi-Terrain Impact Testing and Simulation of a Composite Energy Absorbing Fuselage Section. In: Proceedings of the American helicopter society 60th annual forum, Baltimore, MD, USA, June 2004; pp. 8-10.

7. Block TB. Analysis of the mechanical response of impact loaded composite sandwich structures with focus on foam core shear failure. PhD Thesis, University of Bremen, Germany, 2014.

8. Carstensen T, Cournoyer D, Kunkel E et al. X-cor advanced sandwich core material, In: Proceedings of the 33rd international SAMPE technical conference, Seattle, WA, USA, 2001; pp. 452-466.

9. Kocher C, Watson W, Gomez M et al. Integrity of Sandwich Panels and Beams with Truss-Reinforced Cores. J. Aerosp. Eng. 2002; 15(3), pp.111-117.

10. Rinker M, Zahlen PC, John M et al. Investigation of sandwich crack stop elements under fatigue loading. J Sandwich Struct. Mater. 2011; 14(1), pp.55-73.

11. Kazemahvazi S and Zenkert D. The compressive and shear responses of corrugated hierarchical and foam filled sandwich structures. In: Proceedings of the 8th International Conference on Sandwich Structures, Porto, Portugal, 2008; pp. 275-286.

12. Lascoup B, Aboura Z, Khelil K et al. On the mechanical effect of stitch addition in sandwich panel. Compos. Sci. Technol. 2006; 66, pp. 1385-1398.

13. Zenkert D (ed.). The handbook of sandwich constructions. EMAS Publishing Co. Ltd, UK, 1997.

14. Timoshenko SP. History of the strength of materials. MacGraw-Hill, New York, 1953.

15. Fairbairn W. An account of the construction of the Britania and Conway tabular bridges. John Weale, London, 1849.

16. Vinson JR. Sandwich structures: past, present and future. In: Proceedings of the 7th International conference on sandwich structures, Aalborg, Denmark, 2005; pp. 3–12.

17. Bitzer T. Honeycomb technology: materials, design, manufacturing, applications and testing. Chapman & Hall, London, 1997.

18. Weimer C. Composites in aerospace- future challenges, needs and opportunities. Symposium on the occasion of the 5th anniversary of the institute for carbon composites, Munich, Germany, 2014.

19. ATR product overview, ATR. EWADE - European workshop on aircraft design and education, Naples, Italy, 2011.

20. Wood K. Wind turbine blades: Glass vs. carbon fiber. In: Composite World, case study post. 5/31/2012 (online magazine: https://www.compositesworld.com/articles/wind-turbine-blades-glass-vs-carbon-fiber).

21. Pflug J, Vangrimde B, Verpoest I et al. Continuously produced honeycomb cores. In: Proceedings of the SAMPE 2003, Long Beach, CA, USA, 2003; pp. 602-611.

22. Zahlen PC, Rinker M and Heim C. Advanced manufacturing of large, complex foam core sandwich panels. In: proceedings of the 8th International Conference on Sandwich Structures, Porto, Portugal, 2008; pp. 606-623.

23. Triantafillou TC and Gibson LJ. Failure mode maps for foam core sandwich beams. Mater. Sci. Eng. 1987; 95, pp. 37-53.

24. Product information ROHACELL HERO, February 2014-edition, Evonik Industries AG, Darmstadt, Germany.

25. Foam core impresses in aircraft study. Reinf. Plast. 2014; 52(6), pp. 31-33.

26. Alavi Z, Nia A and Sadeghi M. The effects of foam filling on the compressive response of hexagonal cell aluminum honeycombs under axial loading-experimental study. Mater. Des. 2009; 31, pp. 1216-1230.

27. Wu CL, Weeks CA and Sum TC. Improving honeycomb-core sandwich structures for impact resistance. J. Adv. Mater. 1995; 26(4), pp. 41-47.

28. Vaidya UK, Nelson S, Sinn B et al. Processing and high strain rate impact response of multi-functional sandwich composites. Compos. Struct. 2011; 52, pp. 429-440.

29. Dunker T. Mechanische Charakterisierung von hybriden CFK-Sandwichstrukturen mit einem Kern aus Kork und PET 3D-Core. Master Thesis, University of Bremen, Germany, 2017.

30. Material District, materials catalogue/mikor. Visited on 10/12/2018, https://materialdistrict.com/material/mikor.

31. HP-Textiles, materials catalogue/3D-Core. Visited on 30/05/2018, https://materialdistrict.com/material/mikor.

32. Lascoup B, Aboura Z, Khelil K et al. Core-skin interfacial toughness of stitched sandwich structure. Composites Part B. 2014; 67, pp. 363-370.

33. Lascoup B, Aboura Z, Khelil K et al. Impact response of three-dimensional stitched sandwich composite. Compos. Struct. 2010; 92, pp. 347-353.

34. Henao A, Carrera M, Miravete A et al. Mechanical performance of through-thickness tufted sandwich structures. Compos. Struct. 2010; 92, pp. 2052-2059.

35. Troulis E. Effect of Z-Fiber pinning on the mechanical properties of carbon fibre/epoxy composites. PhD Thesis, Cranfield University, School of Industrial and Manufacturing Science, Advanced Materials, UK, 2003.

36. Marasco AI, Cartié DDR, Parteidge IK et al. Mechanical properties balance in novel Z-pinned sandwich panels: Out-of-plane Properties. Composites Part A. 2006; 37, pp. 295-302.

37. Nanayakkara A, Feith S and Mouritz AP. Experimental analysis of the through-thickness compression properties of z-pinned sandwich composites. Composites Part A. 2011; 42, pp. 1673-1680.

38. Blok LG, Kratz J, Lukaszewicz et al. Improvement of the in-plane crushing response of CFRP sandwich panels by through-thickness reinforcements. Compos. Struct. 2016; 161, pp. 15-22.

39. Yinying Z, Jun X, Mufeng D et al. Experimental study of partially-cured Z-pinned reinforced foam core composites: K-Cor sandwich structures. Chin. J. Aeronaut. 2013; 27(1), pp. 135-159.

40. Rice MC, Fleischer CA and Zupan M. Study on the collapse of pin-reinforced foam sandwich panel cores. Exp. Mech. 2006; 46, pp. 197-204.

41. Du L, Guiqiong J and Tao H. Z-pin reinforcement on the core shear properties of polymer foam sandwich composites. J. Compos. Mater. 2009; 43(3), pp. 289-300.

42. Singh P and La Saponara V. Experimental investigation on performance of angle-stitched sandwich structures. In: Proceedings of the 45th AIAA/ASME/ASCE/AHS/ASC Structures, Structural Dynamics and Materials Conference, Palm Springs, California, USA, 2004.

43. Baral N, Cartié DDR, Partridge IK et al. Improved impact performance of marine sandwich panels using through-thickness reinforcement: experimental results. Composites Part B. 2010; 41, pp. 117-123.

44. Nanayakkara A, Feih S, Mouritz AP. Experimental impact damage study of z-pinned foam core sandwich composite. J. Sandwich Struct. Mater. 2012; 14(4), pp. 269-486.

45. Endres G. Tied Foam Core technology and the achieved impact performance of sandwich constructions based on reinforced foam materials. In: CFK-Valley Stade Convention, Stade, Germany, 2014.

46. John M, Geyer A, Schäuble R et al. Comparing unreinforced and pin-reinforced CFRP/PMI foam core sandwich structures regarding their damage tolerance behaviour. In: Proceedings of the 20th International Conference on Composite Materials, Copenhagen, Denmark, 2015.

47. Weber HJ, Siemetzki M, Endres GC. Reinforcement of cellular material. Patent US20090252917 A1, USA, 2009.

48. Abrate S. Localized impact on sandwich structures with laminated facings. Appl Mech Rev. 1997; 50, pp. 69-82.

49. Chai GB and Zhu S. A review of low-velocity impact on sandwich structures. Proceedings of the Institution of Mechanical Engineers, Part L: Journal of Materials Design and Applications. 2011; 225(4), pp. 207-230.

50. Madenci E and Anderson T. Experimental investigation of low-velocity impact characteristics of sandwich composites. Compos. Struct. 2000; 50, pp. 239-247.

51. Zhou J, Hassan MZ, Guan Z et al. The low velocity impact response of foam-based sandwich panels. Compos. Sci. Technol. 2012; 72, pp. 1781-1790.

52. Leijten J, Bersee Hen, Bergsma OK et al. Experimental study of the low-velocity impact behaviour of primary sandwich structures in aircraft. Composites: Part A. 2009; 40(2), pp. 164-175.

53. Fatt HMS and Park KS. Dynamic models for low-velocity impact damage of composites sandwich panels-Part A: Deformation. Compos. Struct. 2001; 52, pp. 335-351.

54. Malekzadeh K, Khalili MR and Mittal RK. Response of composite sandwich panels with transversely flexible core to low-velocity transverse impact: a new dynamic model. Int. J. Impact Eng. 2007; 34, pp. 522-543.

55. Olsson R. Engineering method for prediction of impact response and damage in sandwich panels. J. Sandwich Struct. Mater. 2002; 4(1), pp. 3-29.

56. Olsson R and Block TB. Criteria for skin rupture and core shear cracking induced by impact on sandwich panels. Compos. Struct. 2015; 125, pp. 81-87.

57. Olsson R. Mass criterion for wave controlled impact response of composite plates. Composites Part A. 2000; 31, pp. 879-887.

58. Anderson T. An investigation of SDOF models for large mass impact on sandwich composites. J. Composites Part B. 2005; 36, pp. 135-142.

59. Zhou DW and Stronge WJ. Low velocity impact denting of HSSA lightweight sandwich panel. Int. J. Mech. Sci. 2006; 48, pp. 1031-1045.
60. Foo CC, Seah LK and Chai GB. A modified energy-balance model to predict low-velocity impact response of sandwich composites. Compos. Struct. 2010; 93, pp. 1385-1393.
61. Olsson R. Closed form prediction of peak load delamination onset under small mass impact. Compos. Struct. 2003; 59, pp. 341-349.
62. Olsson R and McManus HL. Improved theory for contact indentation of sandwich panels. AIAA J. 1996; 34(6), pp. 1238-1244.
63. Arezoo S, Tagarielli VL, Siviour CR et al. Compressive deformation of Rohacell foams: effects of strain rate and temperature. Int. J. Impact Eng. 2013; 51, pp. 50-57.
64. HexFlow RTM6 data sheet, December 2014 Document-no. ITA 065f, Hexcel Corporation, Stamford, USA.
65. Tenax HTS Filamentgarn data sheet, Version 5 30-09-2014, Toho Tenax Europe GmbH Wuppertal, Germany.
66. Torayaca T800H, data sheet, Doc-Nr. CFA-007 Toray Carbon Fibers America INC, Santa Ana, CA, USA.
67. 1383 Yarn, data sheet, Product Bulletin_04_2015, PPG Fiber Glass, Cheswick, PA, USA.
68. Dimassi MA, Brauner C, Herrmann AS, et al. Experimental study of the indentation behaviour of tied foam core sandwich structures, In: Proceedings of the 17th European Conference on Composite Materials, Munich, Germany, 2016.
69. Wu Z, Xiao J, Zeng J et al. Compression performance of integrated 3D composite sandwich structures. J. Sandwich Struct. Mater. 2014; 16(1), pp. 5-21.
70. Wang B, Wu L, Jin X et al. Experimental investigation of 3D sandwich structure with core reinforced by composite columns. Mater. Des. 2010; 31, pp. 158-165.
71. Yalkin HE, Icten BM and Alpyildiz T. Enhanced mechanical performance of foam core sandwich composites with through the thickness reinforced core. Composites Part B. 2015; 79, pp. 383-391.
72. Marasco AL. Analysis and evaluation of mechanical performance of reinforced sandwich structure: X-Cor and K-Cor. PhD Thesis, Cranfield University, UK, 2005.
73. Cartié DD and Fleck NA. The effect of pin reinforcement upon the through-thickness compressive strength of foam-cored sandwich panels. Compos. Sci. Technol. 2003; 63, pp. 2401-2409.
74. DIN 53 291:1982. Standard test method testing of sandwiches; flatwise compression test, Berlin, Beuth Verlag GmbH, 1982.
75. Mouritz AP. Compression properties of z-pinned sandwich composites. J Mater Sci. 2006; 41, pp. 5771-5774.
76. Wu Z, Xiao J, Zeng J et al. Experiments on shear performance of integrated 3D composite sandwich structures. J. Sandwich Struct. Mater. 2014; 16(6), pp. 614-632.
77. DIN 53 294. Standard test method testing of sandwiches; shear test, Berlin, Beuth Verlag GmbH, 1982.
78. Scotch-Weld 9323 B/A data sheet, June 2002, 3m Deutschland Gmbh, Neus, Germany.
79. Hexion MGS LR 385 data sheet, November 2008, Hexion Inc., Columbus, Ohio, USA.

80. Nanayakkara A. Structural properties improvements to aerospace sandwich composites using Z-pins. PhD Thesis, RMIT University, Melbourne, Australia, 2013.

81. Du Y and Jiao GQ. Indentation study of Z-pin reinforced polymer foam core sandwich structures. Composites Part A. 2009; 40, pp. 822-829.

82. Zhang X, Hounslow L and Grassi M. Improvement of low-velocity impact and compression-after-impact performance by z-fibre pinning. Compos. Sci. Technol. 2006; 66, pp. 2785-2794.

83. Cartié DDR, Cox BN and Fleck NA. Mechanisms of crack bridging by composites and metallic rods. Composites Part A. 2004; 35, pp. 1325-1336.

84. Fan X and Xiao-qing W. Study on impact properties of through-thickness stitched foam sandwich composites. Compos. Struct. 2010; 92, pp. 412-421.

85. Han F, Yan Y and Ma J. Experimental study and progressive failure analysis of stitched foam core sandwich composites subjected to low-velocity impact, Polym. Compos. 2016.

86. Gómez-del Rio T, Zaera R, Barbero E et al. Damage in CFRPs due to low velocity impact at low temperature. Composites Part B. 2005; 36, pp. 41-50.

87. John M, Skala T, Schlimper R et al. Dimensional changes in CFRP/PMI foam core sandwich structures. Appl. Compos. Mater. 2013; 20, pp. 601-614.

88. John M. Untersuchungen zu Eigenspannungen in CFK-Schaum-Sandwichstrukturen. PhD Thesis, University of Magdeburg, Germany, 2015.

89. Dimassi MA, John M and Herrmann AS. Investigation of the temperature dependent impact behaviour of pin reinforced foam core sandwich structures. Compos. Struct. 2018; 202, pp. 774-782.

90. John M, Skala t and Schäuble R. Monitoring the residual strain formation in CFRP/PMI foam core sandwich structures, In: Proceedings of the 16th European Conference on Composite Materials, Sevilla, Spain, 2014.

91. Grace I, Pilipchuk V, Ibrahim R et al. Temperature effect on non-stationary compressive loading response of polymethacrylimide solid foam. Compos. Struct. 2012; 94, pp. 3052-3063.

92. Brauner C, Block TB and Herrmann AS. Meso-level manufacturing process simulation of sandwich structures to analyse viscoelastic-dependent residual stresses. J. Compos. Mater. 2011; 46, pp. 783-799.

93. Sánchez-Sáez S, Gómez-del Río, Barbero E et al. Static behaviour of CFRPs at low temperature. Composites Part B. 2002; 33, pp. 383-390.

94. Wessels H and Knothe K. Finite Elemente - eine Einführung für Ingenieure. Springer Vieweg, Berlin, Germany, 2016.

95. Klein B. Grundlagen und Anwendungen der Finite-Element-Methode im Maschinen- und Fahrzeugbau. Springer Vieweg, Wiesbaden, Germany, 2015.

96. Nasdala L. FEM-Formelsammlung Statik und Dynamik. Hintergrundinformationen, Tipps und Tricks. Vieweg+Teubner Verlag, Wiesbaden, Germany, 2010.

97. Abaqus Analysis User's Guide, chapter 24.3: Damage and failure for fiber-reinforced composites. Dassault Systèmes, 2016.

98. Hashin Z. Failure criteria for unidirectional fiber composites. J. Appl. Mech. 1980; 47, pp. 329-334.

99. Hashin Z and Rotem A. A fatigue criterion for fiber-reinforced materials. J. Compos. Mater. 1973; 7, pp. 448-464.

100. Bazant ZP and Oh BH. Crack band theory for fracture of concrete. Mater. Struct. 1983; 16, pp. 155-177.

101. Lapczyk I and Hurtado JA. Progressive damage modelling in fiber-reinforced materials. Composites Part A. 2007; 38(11), pp. 2333-2341.
102. Maimí P, Comanho PP, Mayugo JA et al. A thermodynamically consistant damage model for advanced composites. NASA technical report no. NASA/TM-2006-214282, 2006.
103. eFunda Inc. Crack tip deformation. Visited on 10/12/2018, http://www.efunda.com/formulae/solid_mechanics/fracture_mechanics/fm_lefm_mo des.cfm.
104. González EV, Maimí P, Camanho PP et al. Simulation of drop-weight impact and compression after impact tests on composite laminates. Compos. Struct. 2012; 94, pp. 3364-3378.
105. Lopez CS, Camanho PP, Gürdal Z et al. Low-velocity impact damage on dispersed stacking sequence laminates. Compos. Sci. Technol. 2009; 96, pp. 937-947.
106. Feng D and Aymerich F. Damage prediction in composite sandwich panels subjected to low-velocity impact. Composites Part A. 2013; 52, pp. 12-22.
107. Dogan F, Hadavinia H, Donchev T et al. Delamination of impacted composite structures by cohesive zone interface elements and tiebreak contact. Cent. Eur. J. Eng. 2012; 2(4), pp. 612-626.
108. Comanho PP and Dávila CG. Mixed-mode decohesion finite elements for the simulation of delamination in composite materials. NASA technical report no. NASA/TM-2002-211737, 2002.
109. Turon A, Dávila CG, Camanho PP et al. An engineering solution for mesh size effects in the simulation of delamination using cohesive zone models. Engineering Fracture Mechanics. 2007; 74(10), pp. 1665-1682.
110. Song K, Dávila CG, Rose c et al. Guidelines and parameter selection for the simulation of progressive delamination, In: Proceedings of the Abaqus users´ Conference, Newport, USA, 2008.
111. Park K and Paulino HG. Cohesive zone models: a critical review of traction-separation relationships across fracture surfaces. Appl. Mech. Rev. 2011; 64(6).
112. Abaqus Analysis User´s Guide, chapter 32.5: Cohesive elements. Dassault Systèmes, 2016.
113. Camanho PP, Dávila C and De Moura M. Numerical simulation of mixed-mode progressive delamination in composite materials. J. Compos. Mater. 2003; 37(16), pp. 1415-1438.
114. Harper PW and Hallet SR. Cohesive zone length in numerical simulations of composite delamination. Eng. Fract. Mech. 2008; 75(16), pp. 4774-4792.
115. Corigliano A. Formulation, identification and use of interface models in the numerical analysis of composite delamination. Int. J. Solids Struct. 1993; 30(20), pp. 2779-2811.
116. Alfano and Crisfield. Finite element interface models for the delamination analysis of laminated composites: mechanical and computational issues. Int. J. Numer. Methods Eng. 2001; 50(7), pp. 1701-1736.
117. Abaqus Analysis User´s Guide, chapter 23.3.5: Crushable foam plasticity models. Dassault Systèmes, 2016.
118. Ivañez I, Moure MM, Garcia-Castello SK et al. The oblique impact response of composite sandwich plates. Compos. Struct. 2015; 133, pp. 1127-1136.
119. He Y, Zhang X, Long S et al. Dynamic mechanical behaviour of foam-core composite sandwich structures subjected to low velocity impact. Arch. Appl. Mech. 2016; 86, pp. 1605-1619.

120. Reinhardt B. Touchdown simulation, testing and validation of a marslander demonstrator. Master thesis, University of Bremen, Germany. 2015.

121. Rousse M, Jegley DC, McGowan DM et al. Utilization of the Building-Block Approach in Structural Mechanics Research. In: Proceedings of the 46th AIAA/ASME/ASCE/AHS/ASC/ Structures, Structural Dynamics and Material Conference, Austin, Texas, USA, 2005.

122. Li X, Yan Y, Guo L et al. Effect of strain rate on the mechanical properties of carbon/epoxy composite under quasi-static and dynamic loadings. Polym Test. 2016; 52, pp. 254-264.

123. Zhou Y, Wang Y, Xia Y et al. Tensile behavior of carbon fiber bundles at different strain rates. Mater. Lett. Mech. 2010; 64, pp. 246-248.

124. Zhou Y, Wang Y, Jeelani S et al. Experimental study on tensile behavior of carbon fiber and carbon fiber reinforced aluminum at different strain rate. Appl. Compos. Mater. 2007; 14, pp. 17-31.

125. Gerlach R, Siviour CR, Petrinic N et al. Experimental characterisation and constitutive modelling of RTM-6 resin under impact loading. Polymer. 2008; 49, pp. 2728-2737.

126. Buckley CP, Harding J, Hou JP et al. Deformation of thermosetting resins at impact rates of strain. Part I: experimental study. J. Mech. Phys. Solids. 2001; 49, pp. 1517-1538.

127. Gilat A, Goldberg RK and Roberts GD. Strain rate sensitivity of epoxy resin in tensile and shear loading. J. Aerosp. Eng. 2007; 20(2).

128. Ochola RO, Marcus K, Nurick GN et al. Mechanical behaviour of glass and carbon fibre reinforced composites at varying strain rates. Compos. Struct. 2004; 63, pp. 455-467.

129. Shokrieh MM and Omidi JM. Tension behavior of unidirectional glass/epoxy composites under different strain rates. Compos. Struct. 2009; 88, pp. 595-601.

130. Gurusideswar S, Srinivasan N, Velmurugan R et al. Tensile response of epoxy and glass/epoxy composites at low and medium strain rate regimes. Procedia. Eng. 2017; 173, pp. 686-693.

131. Taniguchi N, Nishiwaki T and Kawada H. Tensile strength of unidirectional CFRP laminate under high strain rate. Adv. Compos. Mater. 2007; 16(2), pp. 167-180.

132. Gilat A, Goldberg RK and Roberts GD. Experimental study of strain-rate-dependent behaviour of carbon/epoxy composite. Compos. Sci. Technol. 2002; 62, pp. 1469-1476.

133. Koerber H, Xavier J and Camanho PP. High strain rate characterisation of unidirectional carbon-epoxy IM7-8552 in transverse compression and in-plane shear using digital correlation. Mech. Mater. 2010; 42, pp. 1004-1019.

134. Hsiao HM and Daniel IM. Strain rate behavior of composite materials. Composites Part B. 1998; 29(5), pp. 521-533.

135. Arezoo S, Tagarielli VL, Petrinic N et al. The mechanical response of Rohacell foams at different length scales. J. Mater Sci. 2011; 46, pp. 6863-6870.

136. Williams DA and Lopez-Anido RA. Strain rate and temperature effects of polymer foam core material. J. Sandwich Struct. Mater. 2014; 16(1), pp. 66-87.

137. Mahfuz H, Thomas T, Rangari V et al. On the dynamic response of sandwich composite and their core materials. Compos. Sci. 2006; 66, pp. 2465-2472.

138. Mahfuz H, Al Mamun W, Haque A et al. An innovative technique for measuring the high strain rate response of sandwich composites. Compos. Struct. 2000; 50, pp. 279-285.

139. Saha MC, Mahfuz H, Chakravaty UK et al. Effect of density, microstructure, and strain rate on compression behaviour of polymeric foams. Mater. Sci. Eng., A. 2005; 406, pp. 328-336.

140. Luong DD, Pinisetty D and Gupta N. Compressive properties of closed-cell polyvinyl-chloride foams at low and high strain rates: Experimental investigation and critical review of state of the art. Composites Part B. 2013; 44(1), pp. 403-416.

141. DIN EN ISO 14125. Standard test method, testing of fibre-reinforced plastic composites: Determination of flexural properties, Berlin, Beuth Verlag GmbH, 2011.

142. DIN EN ISO 527-4. Standard test method, testing of fibre-reinforced plastic composites: Determination of tensile properties, Berlin, Beuth Verlag GmbH, 1997.

143. Deshpande VS and Fleck NA. Isotropic constitutive model for metallic foams. J. Mech. Phys. Solids. 2000; 48, pp. 1253-1276.

144. Dimassi MA, Brauner C and Herrmann AS. Experimental study of the mechanical behaviour of pin reinforced foam core sandwich materials under shear load, In: Proceedings of the 18th Chemnitz seminar on materials engineering, Chemnitz, Germany, 2016; 59, pp. 268-275.

145. Xia F, Wu XQ and Li JL. Numerical simulation of impact responses on through-thickness stitched foam core sandwich composite. Appl. Compos. Mater. 2013; 20(6), pp. 1041-1054.

146. Dimassi MA, Dunker T, Brauner C et al. Improvement of the impact behaviour of foam core sandwich through the use of a cork layer as impact shield, In: Proceedings of the 12th international conference on sandwich structures (ICSS-12), Lausanne, Switzerland, 2018, pp. 270-272.

This PhD thesis contains results that were generated in the following supervised student work:
S. Warhus, (2015), Master Thesis, University of Bremen, Experimentelle Kennwertermittlung und numerische Simulation des Crushing-Verhaltens von Pin-verstärkten Sandwichstrukturen.

List of figures

List of tables

A Appendix

A.1 Calculation of effective stiffnesses of sandwich panels

Effective bending stiffness of the skin Q_f^* (subscript f for the face sheet):

$Q_f^* = 12 D_f^* / h_f^3$, where h_f is the thickness of the face sheet.

$D_f^* \approx \sqrt{D_{f11} D_{f22}(\eta + 1)/2}$, where $\eta = (D_{f12} + 2 D_{f66})/\sqrt{D_{f11} D_{f22}}$, and D_{fij} are the element of the bending stiffness matrix of the skin.

$D_{f11} = \frac{1}{12} E_{fbx} h_f^3 / \psi_{fb}$, where b means bending properties of the skin.

$D_{f22} = \frac{1}{12} E_{fby} h_f^3 / \psi_{fb}$

$D_{f66} = \frac{1}{12} G_{fbxy} h_f^3$

$D_{f12} = D_{f22} \vartheta_{fbxy}$

$\psi_{fb} = \left(1 - \vartheta_{fbxy} \vartheta_{fbyx}\right) = (1 - \frac{\vartheta_{fbxy}^2 E_{fby}}{E_{fbx}})$

Effective bending stiffness of the core Q_{cz}^* (subscript c for the core):

$Q_{cz}^* = E_c/(1 - \vartheta_c^2)$ for isotropic foams, and

$Q_{cz}^* = E_{cz}$ for honeycomb, where E_c and ϑ_c is are the Young's modulus and Poisson's ratio of the foam core and E_{cz} is the Young's modulus of the honeycomb core in out-of-plane direction.

A.2 Results of the tensile test of the CFRP face sheet

Specimens layups:

- Triax-0°: $(45/0/-45)_4(-45/0/45)_4$
- Triax-90°: $(45/90/-45)_4(-45/90/45)_4$

5 specimens per layup were tested in accordance to DIN EN ISO 527-4

Average Test results:

Triax-0°:

Average material properties:

FVC [%]	Width [mm]	Thickness [mm]	Max. Stress [MPa]	Tensile modulus [MPa]
57.8	25.11	3.28	860.72	50283.24

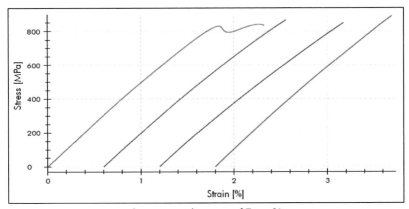

Stress-strain diagrams of Triax-0°

Triax-90°:

Average material properties:

FVC [%]	Width [mm]	Thickness [mm]	Max. Stress [MPa]	Tensile modulus [MPa]
57.8	25.11	3.28	282.91	20551.16

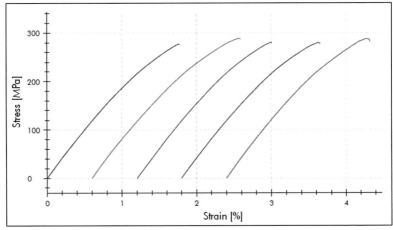

Stress-strain diagrams of Triax-90°

A.3 Validation of the sandwich impact model

Specimen with 10 mm core and 1.5 mm face sheets at 35 J impact energy and 3.7 ms-1 impact velocity:

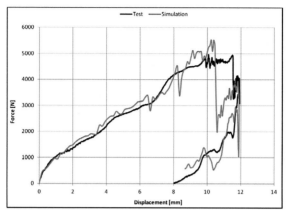

Experiment vs. Simulation: Force- displacement curves

Specimen with 10 mm core and 1.5 mm face sheets at 35 J impact energy and 4.77 ms-1 impact velocity:

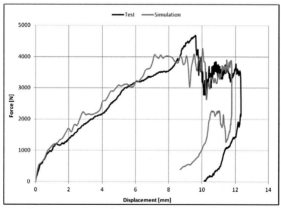

Experiment vs. Simulation: Force- displacement curves

Specimen with 10 mm core and 1.5 mm face sheets at 50 J impact energy and 4.44 ms⁻¹ impact velocity:

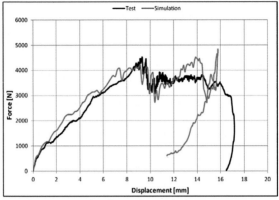

Experiment vs. Simulation: Force- displacement curves

Specimen with 10 mm core and 1.5 mm face sheets at 20 J impact energy and 3.7 ms⁻¹ impact velocity:

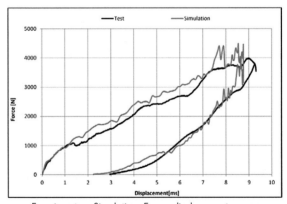

Experiment vs. Simulation: Force- displacement curves

Specimen with 16.3 mm core and 1.5 mm face sheets at 50 J impact energy and 4.44 ms⁻¹ impact velocity:

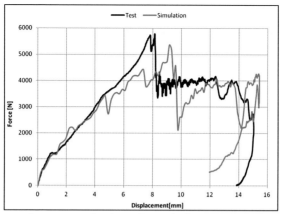

Experiment vs. Simulation: Force- displacement curves

Specimen with 16.3 mm core and 2.25 mm face sheets at 35 J impact energy and 4.77 ms⁻¹ impact velocity:

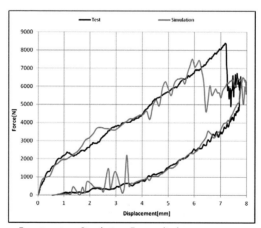

Experiment vs. Simulation: Force- displacement curves

Bisher erschienene Bände der Reihe

Science-Report aus dem Faserinstitut Bremen

ISSN 1611-3861

Alle erschienenen Bücher können unter der angegebenen ISBN-Nummer direkt online (http://www.logos-verlag.de) oder per Fax (030 - 42 85 10 92) beim Logos Verlag Berlin bestellt werden.